Satish Kumar Peddapelli, Sridhar Gaddam

Electrical Machines

Also of interest

Electrical Engineering
Viktor Hacker, Christof Sumereder, 2020
ISBN 978-3-11-052102-3, e-ISBN (PDF) 978-3-11-052111-5,
e-ISBN (EPUB) 978-3-11-052113-9

Protecting Electrical Equipment
Vladimir Gurevich, 2019
ISBN 978-3-11-063596-6, e-ISBN (PDF) 978-3-11-063928-5,
e-ISBN (EPUB) 978-3-11-063606-2

Systems, Automation, and Control
Edited by Nabil Derbel, Faouzi Derbel and Olfa Kanoun, 2020
ISBN 978-3-11-059024-1, e-ISBN (PDF) 978-3-11-059172-9,
e-ISBN (EPUB) 978-3-11-059031-9

Electrochemical Energy Systems
Artur Braun, 2018
ISBN 978-3-11-056182-1, e-ISBN (PDF) 978-3-11-056183-8,
e-ISBN (EPUB) 978-3-11-056195-1

Satish Kumar Peddapelli,
Sridhar Gaddam

Electrical Machines

A Practical Approach

DE GRUYTER

Authors
Dr. Satish Kumar Peddapelli
Associate Professor
Department of Electrical Engineering
University College of Engineering, Osmania University
Hyderabad, Telangana State, India

Dr. Sridhar Gaddam
Associate Professor
Department of Electrical and Electronics Engineering
Jyothishmathi Institute of Technology and Science
Karimnagar, Telangana State, India

ISBN 978-3-11-068195-6
e-ISBN (PDF) 978-3-11-068227-4
e-ISBN (EPUB) 978-3-11-068244-1

Library of Congress Control Number: 2020933280

Bibliographic information published by the Deutsche Nationalbibliothek
The Deutsche Nationalbibliothek lists this publication in the Deutsche Nationalbibliografie; detailed bibliographic data are available on the Internet at http://dnb.dnb.de.

Cover image: Bosca78/ E+/Getty Images
Typesetting: Integra Software Services Pvt. Ltd.
Printing and binding: CPI books GmbH, Leck

www.degruyter.com

Dedicated to all my teachers

Preface

Electrical Machines – *A Practical Approach* is written to present the concepts of electrical machines in a practical manner for the students of under graduate and post graduate. There are many textbooks available on electrical machines that discuss the theoretical concepts, which are difficult for the students to understand. This book presents the theoretical, mathematical and practical concepts of the most commonly used electrical machines in way that is easily understandable to the readers. It provides a step by step procedure to obtain the performance characteristics of electrical machines on various operating, testing and load conditions. The necessary experiments on electrical machines have been conducted at rated conditions and the procedure, observations, calculations and results have been presented in a systematic manner which includes neat circuit diagrams, performance curves and graphs/plots.

The objective and viva-voce questions are included in order to equip the students to compete in various competitive and practical examinations. The solved numerical problems and exercise problems are also provided at the end of each chapter.

The contents of this book are methodically organized in six chapters; Chapter 1 and Chapter 2 are on DC generators and DC motors, respectively, followed by transformers in Chapter 3, induction motors in Chapter 4 and synchronous generators and synchronous motors in Chapter 5 and Chapter 6, respectively.

The authors welcome the suggestions and additions for the improvement of this book.

https://doi.org/10.1515/9783110682274-202

Contents

Preface —— VII

1 DC Generators —— 1

2 DC Motors —— 38

3 Transformers —— 87

4 Induction Motors —— 123

5 Synchronous Generators —— 150

6 Synchronous Motor —— 172

Bibliography —— 187

Index —— 189

1 DC Generators

1.1 Introduction

A machine that changes mechanical power into electrical power by using the principle of magnetic induction is called generator. Whenever a conductor cuts the magnetic field, an emf(Electromotive Force) is generated in that conductor. The magnitude of the generated emf is proportional to $d\phi/dt$ and the polarity depends on the direction of the flux and the conductor. Fleming's right hand rule is used to determine the current direction.

1.2 Operation of DC generator

The construction of a simple DC generator is shown in Fig.1.1. Whenever a conductor is kept in a magnetic field, an emf is induced in the conductor. In a DC generator, magnetic field is generated by field coils and the armature conductors are rotated in the magnetic field. Thus, an emf is produced in the armature conductors. The voltage/current output waveform is shown in Fig. 1.2.

1.3 Voltage equation of a generator

Let

P = Number of poles

Φ = Flux/pole in weber

Z = Number of conductors (Armature) = Number of slots × Number of conductors/slot

N = No. of rotations in rpm; A = No. of parallel paths

E_g = emf generated in any one parallel path in the armature

$$\text{Average voltage produced per conductor} = \frac{d\phi}{dt}\ \text{Volt} \tag{1.1}$$

The change in flux per conductor in one revolution $d\phi = \phi \times P$ webers

$$\text{The flux changes}\ \frac{N}{60}\ \text{rotations/sec}$$

Therefore, the time taken to complete one complete rotation is $dt = \frac{60}{N}$ second

$$\text{Voltage generated/conductor}(E_g) = \frac{d\phi}{dt} = \frac{\phi P N}{60}\ \text{webers} \tag{1.2}$$

https://doi.org/10.1515/9783110682274-001

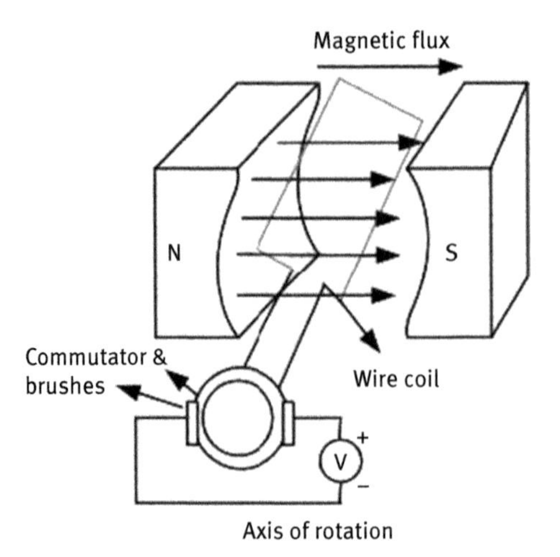

Fig. 1.1: Principle of the generator.

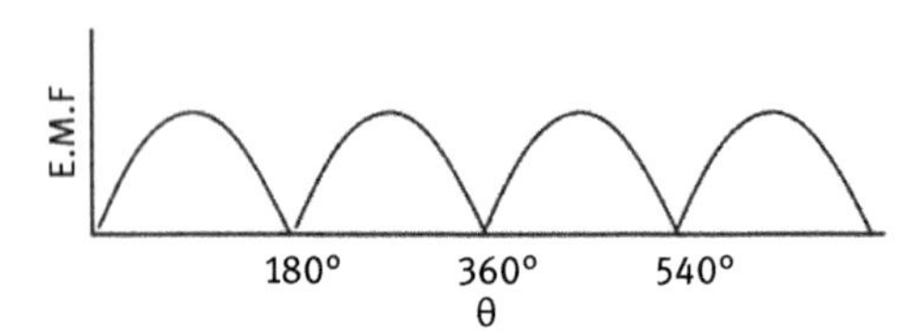

Fig. 1.2: Voltage across the load.

For a simple wave-wound generator

$$A = 2;$$

The number of conductors per path $= \frac{Z}{2}$

$$E_g/\text{path} = \frac{\phi\, P\, N}{60} \times \frac{Z}{2} = \frac{\phi\, Z\, N\, P}{120}\,\text{Volt} \tag{1.3}$$

For a simple lap-wound generator

$$A = P;$$

The number of conductors per path $= \frac{Z}{P}$

$$\text{Voltage generated/conductor (Eg)} = \frac{\phi PN}{60} \times \frac{Z}{P} = \frac{\phi ZN}{60} \tag{1.4}$$

1.4 Types of generators

They are divided into two types based on the energization of field windings, namely separately-excited and self-excited generators

(a) Separately-excited generators

In this type of generator, the field winding is energized separately by a DC voltage source as shown in Fig. 1.3.

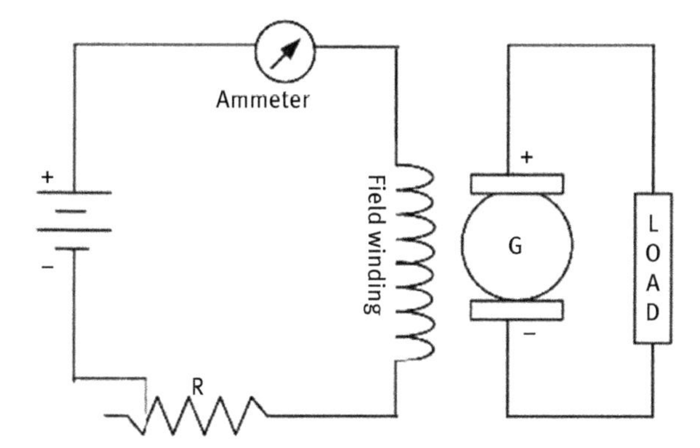

Fig. 1.3: Separately-excited DC generator.

(b) Self-excited generators

The field winding is excited by the voltage generated by them. When the armature conductors are rotated, a small voltage is produced due to residual magnetic field and allows a small current through field winding which makes the residual flux stronger.

The self-excited generators are categorized based on the connection of field windings to the armature winding such as (i) shunt, (ii) series and (iii) compound

(i) Shunt generators

In shunt generator, the armature and field windings are connected in parallel. Rated generated emf is applied across the shunt winding. It is shown in Fig. 1.4(a).

$$\text{Equation of the generator}(E_g) = \text{Terminal Voltage} + \text{Voltage drop in the armature winding} = V + I_a R_a \tag{1.5}$$

From the Fig. 1.4

$$I_a = I_{sh} + I_L \tag{1.6}$$

(ii) Series generator

In series generator, the field and armature windings are connected in series and both the windings carries full load current. It is shown in Fig. 1.4(b).

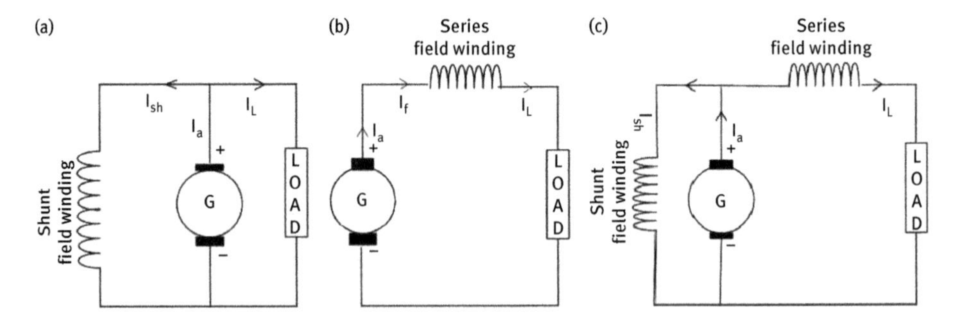

Fig. 1.4: Types self-excited DC generators.

$$\begin{aligned} \text{Equation of the Generator}(E_g) &= \text{TerminalVoltage} \\ &\quad + \text{Voltage drop in the armature winding} \\ &= V + I_a\,(R_{se} + R_a) \end{aligned} \tag{1.7}$$

and $I_a = I_{sh} = I_L$

(iii) Compound generator

In this, the series and shunt windings are connected as shown in Fig. 1.4(c) and the shunt field is stronger than series field. If the series and shunt field flux adds, then the generator is said to be commutatively compounded and if both field fluxes are subtracted, the generator is said to be differentially compounded.

1.5 Armature reaction

The distribution of the main flux developed by the main poles of a generator is affected by another magnetic flux developed by the armature current. This is referred to as armature reaction.

The two effects of armature reaction are (i) demagnetizing effect and (ii) cross magnetizing effect. The generated voltage is reduced due to the demagnetizing effect, and cross magnetizing effect creates sparking at the brushes.

The effects of armature reaction are explained using Fig. 1.5. The axis along which no voltage is developed in the armature conductors is called magnetic neutral axis and these move in parallel with the flux lines. The brushes are positioned along Magnetic Neutral Axis (MNA).

The magnetic neutral axis moves in the same direction of the generator rotation and moves in opposite direction of the motor rotation. The shift between Geometrical Neutral Axis (GNA) and MNA depends on the magnitude of the current flowing through the armature conductors i.e. load current of the machine.

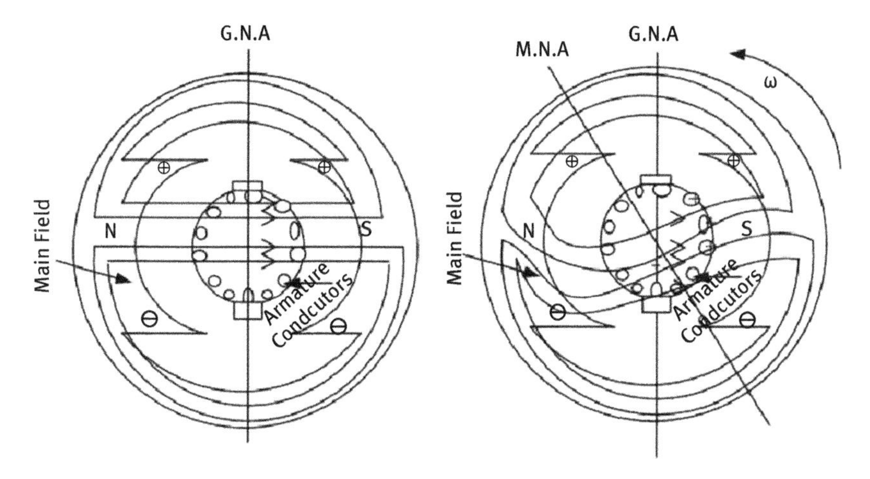

Fig. 1.5: Armature reaction.

1.6 Characteristics of DC generator

The following characteristics are useful to analyses the DC generator.

a) No-load or magnetization characteristic (E_0 versus I_f)
The magnetization characteristic curve is drawn between the voltage generated under no-load condition of the generator armature conductors (E_0) and field current at constant speed.

b) Internal characteristic (E_0 versus I_a)
The internal characteristics curve of the generator is drawn between the voltage generated under no-load condition of the generator armature conductors (E_0) and actually induced in the armature and the current flowing through armature (I_0)

c) External Characteristic (V_L versus I_L)
These characteristics explain the information between the voltage available across the output terminals (V_L) and the current flowing through the load (I_L). The external characteristic curve lies below the internal characteristic curve due to the voltage drop in the armature resistance. The terminal voltage V_L is calculated by subtracting I_a R_a drop from no-load generated voltage E_0.

1.6.1 Open–circuit characteristic for shunt generator

The magnetization or open circuit characteristic (OCC) or no-load characteristics are obtained for shunt or series-connected generators. The generator field winding is excited

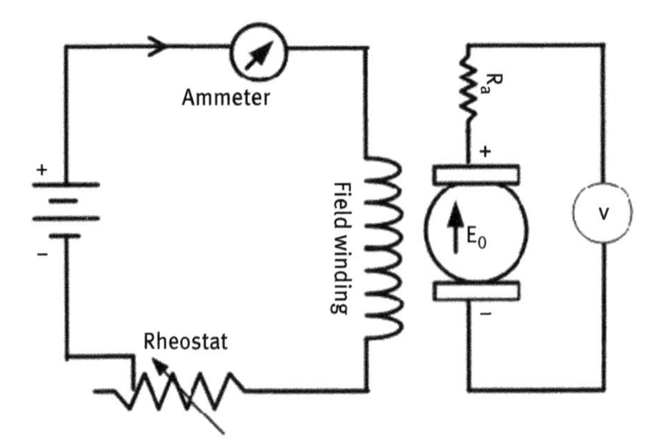

Fig. 1.6: Self-excited generator.

separately from an external DC voltage source as shown in Fig. 1.6. The field/exciting current is varied using the rheostat connected in series with the field winding and its value is indicated on the ammeter (A).

Using prime mover, the machine is running at constant speed without load and the generator produces the voltage which is determined by the voltmeter attached between the armature terminals of the machine.

The generated voltage under no-load condition is recorded by increasing the field current by appropriate values and the plot is drawn between field current verses (I_f) and no-load generated voltage(E_0) as shown in Fig. 1.7. Even if field current is zero some voltage is generated due to the presence of residual magnetism. Therefore, the curve begins from point "p", the initial portion is a straight line after that the curve has curvature due magnetic inertia.

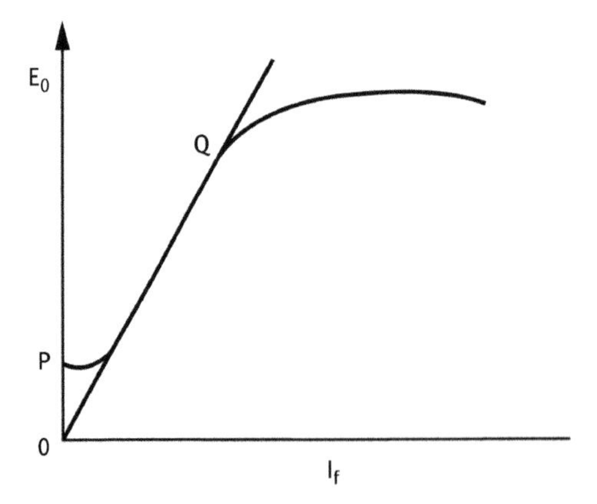

Fig. 1.7: Open circuit characteristic (OCC) curve.

The shunt generator does not start build up voltage, if the poles do not have some residual magnetism and its resistance should be minimum than the critical value. The real value depends on resistance of the generator and rotor speed. The shunt generator magnetization characteristic is shown in Fig. 1.8 and is drawn by increasing the field by maintaining constant speed. From the curve, the generated voltage is PQ for a shunt current OA, and OP is voltage due to residual magnetism.

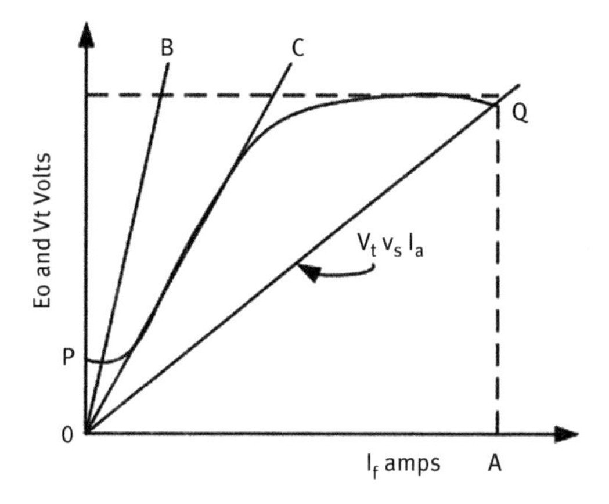

Fig. 1.8: Magnetization Characteristics of a shunt generator.

$$\text{Corresponding resistance of shunt circuit} = \frac{\text{Terminal Voltage}}{\text{Shunt Current}} = \frac{AQ}{OA}$$
$$= \tan(QOA) = \text{Slope of } OQ$$

The OQ represents resistance line. The machine will produce highest voltage AQ for small resistance and if the resistance of the field is increased, the highest generated voltage will decrease at a given speed.

The line OB does not cut the OCC curve due to increased field resistance, therefore the generator fails to excite, hence no voltage is generated. The critical field resistance is defined as the tangent of the magnetization curve at a specified speed.

1.6.2 External characteristic of a shunt generator

To find external characteristic, a load is connected across the terminals of the generator under running condition. When the generator is loaded, it is observed that the terminal voltage decreases with increase in load. The drop in terminal voltage is objectionable particularly when the generator is delivering current for a load and the power which is required for a certain load should remain almost stable and should not depend on the load.

The terminal voltage drop for a generator when loaded is due to the following reasons:

i) Voltage drop in the armature resistance.
ii) Effect of the armature reaction.
iii) Reduction in field current due to decreased terminal voltage by the resistance of the armature and armature reaction effects.

$$\text{The generator terminal voltage } V = E_o - I_a R_a ; \quad E_o = K\phi N \tag{1.8}$$

To draw the external characteristics of the generator, first it is running on no-load such that the generator delivers rated voltage. It is shown in Fig. 1.9 as "0–a." The load current and terminal voltage is measured by increasing the load by appropriate steps with the field current constant. The external characteristics of Fig. 1.9 are plotted by using the recorded values.

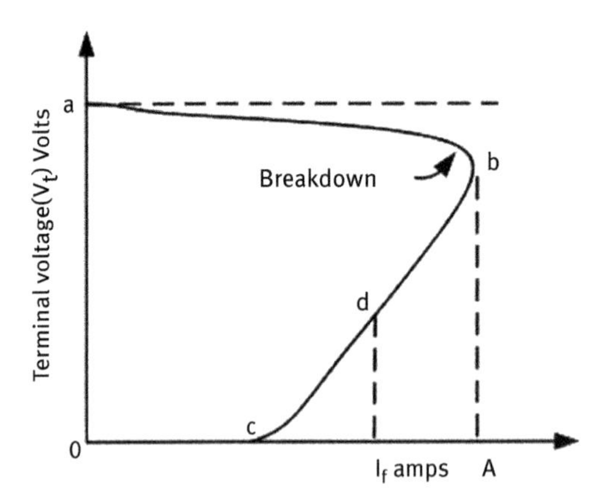

Fig. 1.9: External characteristics.

The section "a–b" is the effective section of the external characteristics. After this, if the current supplied to the load is increased, a little extra voltage drop is added and this condition is valid up to point "b" is achieved and this point is called breakdown point. Further any raise in load current, the terminal voltage will reduce very sharply ("b–c").

1.6.3 Internal characteristic of a shunt generator

This characteristic is obtained by plotting the curve between no-load voltage (E) and armature current.

For shunt generator,

$$I_a = I_f + I_L \tag{1.9}$$

$$E_a = V + I_a R_a \tag{1.10}$$

$$I_f = \frac{V}{R_f} \tag{1.11}$$

In Fig. 1.10 the curve "a–b" corresponds to the external characteristics. The voltage drop due to armature resistance (0M) is drawn and drop due the brush contact is assumed constant.

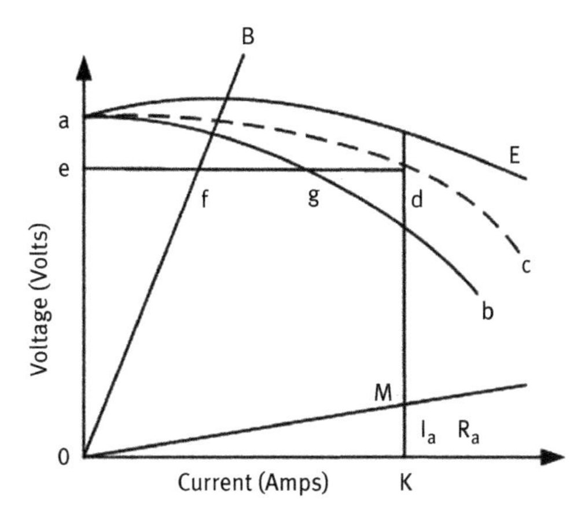

Fig. 1.10: Internal characteristics.

If brush contact resistance is assumed constant, then armature voltage drop is proportional to the armature current. In the plot 0K is the armature current and corresponding armature voltage drop is $I_a R_a$ = MK. The internal characteristic "a–c" is obtained by combing the brush contact drop and $I_a R_a$ drop.

1.6.4 Characteristics of series generator

The DC series generator magnetization characteristics appears similar to the shunt generator.

The terminal voltage is very small at no load due to zero field current and the small voltage is generated because of residual field flux. The generated voltage increases quickly as the load current increases but initially the generated voltage raises more rapidly than $I_a(R_a+R_f)$. As the $I_a(R_a+R_f)$ drop rises, the terminal voltage increases and reaches to saturation then the generated voltage becomes almost constant.

$$I_a = I_f = I_L \tag{1.12}$$

$$V = E - I_a R_a - I_f R_f = E - I_a (R_a + R_f) \tag{1.13}$$

The series generator characteristics are shown in Fig. 1.11. It is noticeable that the series generator is not a constant-voltage source. Therefore, its voltage regulation is negative.

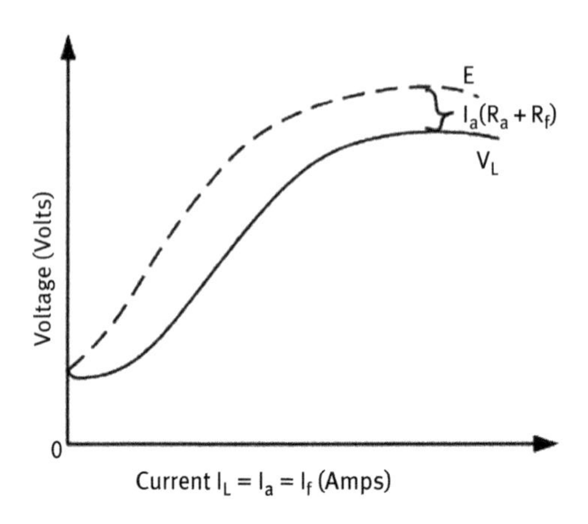

Fig. 1.11: Magnetization curve of a series DC generator.

1.6.5 Compound wound generator

In level compound wound generator, the no-load and full load voltages are equal but the original characteristic curve is not flat due to nonlinearity in the demagnetizing effects of armature reaction. In over compound wound generator, rated load voltage is superior to the no-load voltage otherwise called under compound wound generator. These characteristics are shown in Fig. 1.12.

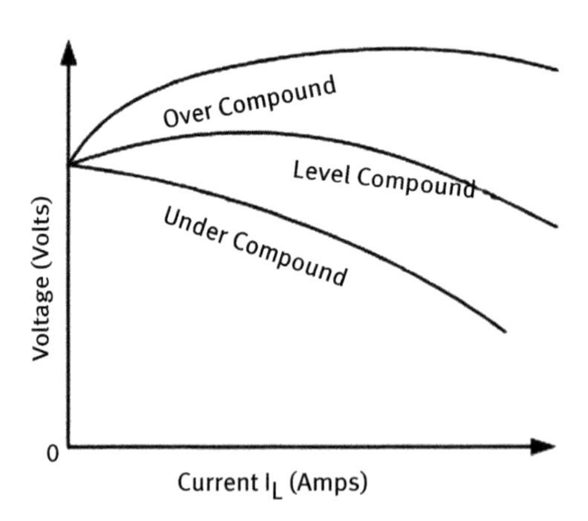

Fig. 1.12: Characteristics of compound generator.

1.7 Testing of DC generators

1.7.1 Magnetization/no-load/open circuit characteristics of DC shunt generator

The critical field resistance of DC shunt generator is evaluated practically in laboratory by adopting the following method. The list of apparatus required is shown in Table 1.1.

1.7.1.1 Apparatus required

Table 1.1: List of apparatus required.

S.No	Name of the equipment	Range	Quantity
1	Ammeter	0–2A(MC)	1
2	Voltmeter	0–500V(MC)	1
3	Rheostat	400 Ω /1.7A	1
4	Rheostat	1,000 Ω /1.7A	2
5	Tachometer	0–2,000 rpm	1

1.7.1.2 Theory

The plot between the field current and generated voltage is called the magnetization characteristics.

In a DC generator, the generated voltage at a given speed is written as

$$E_g = \frac{\phi NPZ}{60A} \text{ Volts} \tag{1.14}$$

Open–circuit characteristics

Gradually increase the field current from zero to its rated value by driving the generator at rated speed and is maintained constant. The terminal voltage (V) and corresponding field current (I_f) of the generator is recorded. At no-load the V and E_g are equal due to small armature voltage drop. Thus, the OCC and the magnetization curve is the same.

Critical field resistance (R_C)

The maximum field resistance at which the generator just excites is called critical field resistance and if the field resistance (R_f) is increased further the generator does not excite and no voltage is generated by the generator.

1.7.1.3 Circuit diagram

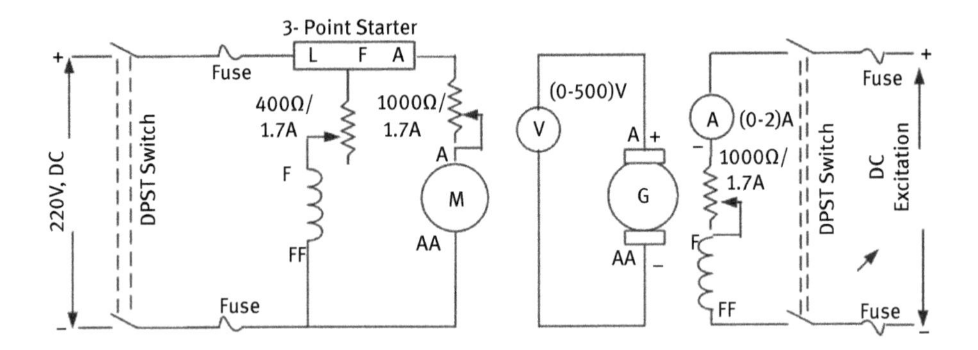

Fig. 1.13: DC shunt generator circuit diagram for magnetization characteristics.

Shunt field resistance (R_{sh})

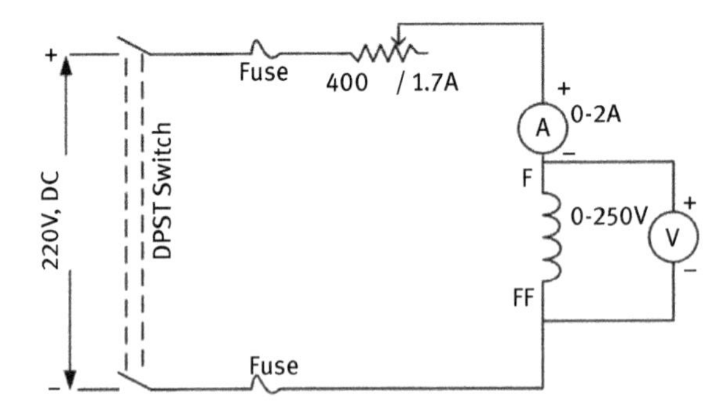

Fig. 1.14: DC Shunt generator circuit diagram: evaluation of critical field resistance.

1.7.1.4 Procedure

1. The name plate details of the shunt motor and generator should be noted.
2. The connections of various terminals are connected as per the circuit with DPST off position.
3. Before starting the machine, the field rheostat of the generator and motor should keep at minimum resistance position and the motor armature rheostat position must be minimum.
4. Switch on the supply (220V DC) and using 3-point starter get the motor speed to its rated value by adjusting armature and field rheostat. Note down the residual voltage of the generator
5. Now switch on the generator side DPST and apply DC excitation voltage to the generator field.

6. Increase the field current of the generator by adjusting generator field, note down the values of I_f and E_g up to 125% of the generator rated voltage.
7. The speed of the generator should be maintained constant throughout the experiment.
8. Draw the magnetization plot between Eg verses I_f and calculate the critical field resistance.

1.7.1.5 Observation table

Table 1.2: Observation table.

S.No	Field current I_f (A)	Generated voltage E_g (V)	Speed (rpm)
1			
2			
3			
4			

1.7.1.6 Model graph

1. Generated emf (E_g) versus field current (I_f)

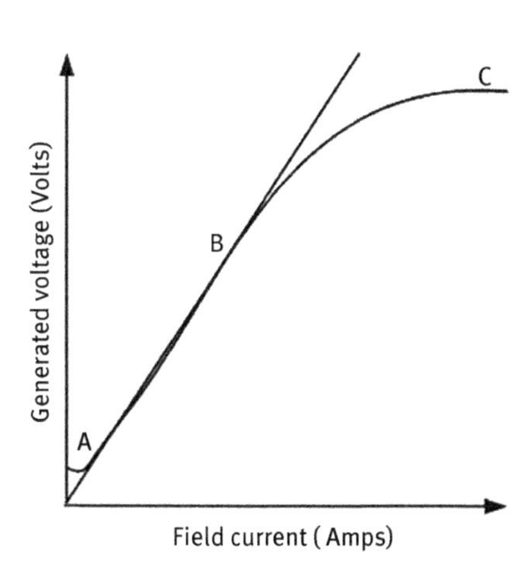

Fig. 1.15: Model graph of E_g versus I_f curve of DC shunt generator for evaluation of critical field resistance.

1.7.1.7 Practical calculations

The critical field resistance for DC shunt generator with following ratings and readings were tabulated.

Table 1.3: Machine ratings.

	Motor	Generator
Rated Power	3.5 kW	3.5 kW
Rated Speed	1,500 rpm	1,500 rpm
Full Load Current	21 A	16 A
Excitation Current	0.5 A	0.9 A

1.7.1.8 Observation table

Table 1.4: Practical observations.

S.no	Current through the field I_f (A)	Voltage generated by the generator E_g(V)	$R_f = Eg/I_f$ (Ω)
1	0	60	600
2	0.2	105	525
3	0.3	145	483.33
4	0.4	170	425
5	0.5	190	380
6	0.6	210	350
7	0.7	220	314.28
8	0.8	240	300
9	0.9	245	272.22
10	1.0	250	250
11	1.1	255	231.81
12	1.2	265	220.83

1.7.1.9 Calculations

$$\text{Field Resistance } R_f = \frac{\text{Generated emf}}{\text{Field Current}}\ \Omega$$

I) $R_f = \frac{60}{0.1} = 600\Omega$

II) $R_f = \frac{105}{0.2} = 525\Omega$

III) $R_f = \dfrac{145}{0.3} = 483\Omega$

IV) $R_f = \dfrac{170}{0.4} = 425\Omega$

V) $R_f = \dfrac{190}{0.5} = 380\Omega$

VI) $R_f = \dfrac{210}{0.6} = 350\Omega$

1.7.1.10 Conclusion

The shunt generator magnetization characteristics are drawn and shown in above Fig. 1.16 and the critical resistance (R_c) is calculated and its value is 600 Ω.

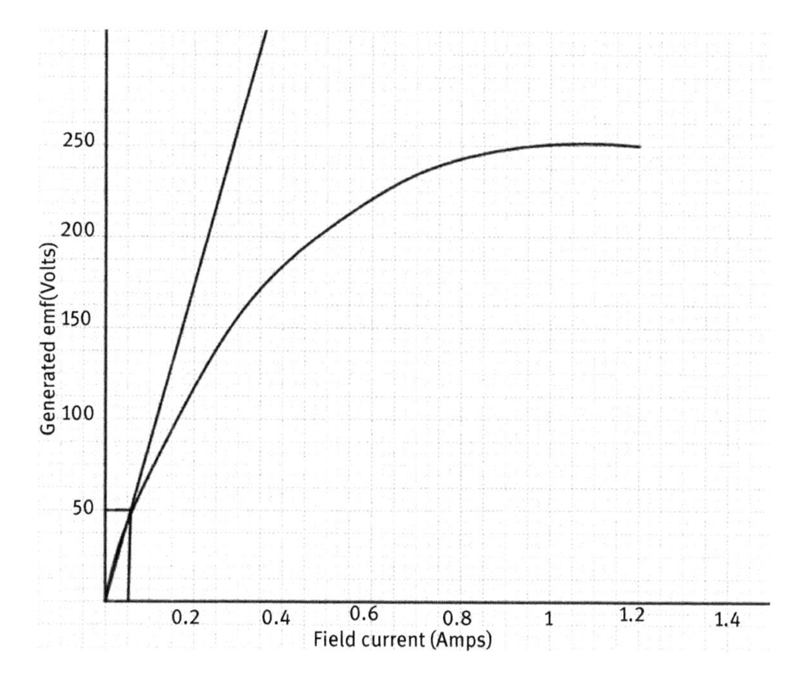

Fig. 1.16: Graph between generated emf, E_g versus field current and I_f curve of DC shunt generator.

1.7.2 Testing of DC shunt generator – load test

The internal and external characteristics of DC shunt generator are evaluated by conducting load test practically in laboratory by adopting the following method.

1.7.2.1 Apparatus required

Table 1.5: List of apparatus required.

Name of the equipment	Range	Quantity
Ammeter	0–20 A(MC)	1
Ammeter	0–5 A(MC)	1
Voltmeter	0–300 V(MC)	1
Voltmeter	0–3 V(MC)	1
Rheostat	400 Ω /1.7 A	2
Rheostat	100 Ω /5 A	1
Resistive Load	0–3,000 W	1
Tachometer	0–2,000 rpm	1

1.7.2.2 Theory

The DC shunt generator performance is analyzed by the following characteristics:

i) External characteristics

These characteristics are obtained by drawing the curve between load voltage and current at a specified speed of the generator. The behavior of this characteristic is well understood by the following relations.

$$I_f = V_f/R_f \text{ (Ohm's law)} \tag{1.15}$$

$$I_L = V_L/R_L \text{ (Ohm'slaw)} \tag{1.16}$$

$$I_a = I_f + I \text{ (KCL)} \tag{1.17}$$

$$E_g = V_L + I_a R_a \text{ (KVL)} \tag{1.18}$$

The voltage (E_g) is generated by the resultant air gap flux which is present in the air gap due to the flux produced by the generator field current and armature current (I_f and I_a). At no-load, the load current I_L is zero, and it starts increasing as load increases. Hence the armature voltage drop I_aR_a raises and terminal voltage V_L of the generator is reduced according to the eq. 1.18. The generated voltage E_g is lowered by itself due to increased effect of armature reaction and reduction in terminal voltage V_L. These effects are shown in Fig. 1.19 and is known as external characteristic.

ii) Internal characteristics

These characteristics are obtained by drawing a curve between voltage generated by the armature conductor and armature current at a specified speed of the generator.

$$I_a = I_L + I_f \tag{1.19}$$

$$E = V_L + I_a R_a \tag{1.20}$$

1.7.2.3 Circuit diagram

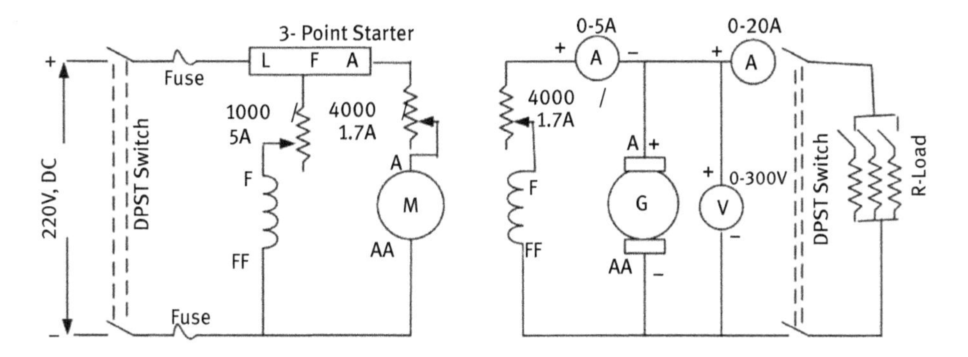

Fig. 1.17: Circuit diagram of DC shunt generator: load test.

Armature resistance (*R*a)

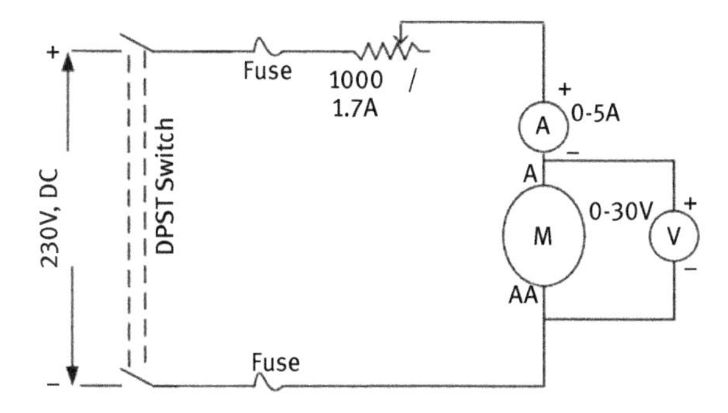

Fig. 1.18: Circuit diagram of DC shunt generator: armature resistance evaluation.

1.7.2.4 Procedure

1. The name plate details of the shunt motor and generator should be noted.
2. The connections of various terminals are connected as per the circuit with DPST off position.

3. Set the generator field current at minimum value by the rheostat connected in the field circuit and adjust the motor field current to rated value gradually by the rheostat connected in the field circuit.
4. Switch off all loads.
5. Switch on the supply (220 V DC) and using 3-point starter get the motor speed to its rated value by adjusting armature and field rheostat. Note down the residual voltage of the generator.
6. With switch off on the generator side DPST, the generator field current is adjusted by its rheostat till the rated voltage of the generator is produced.
7. Now close the generator side DPST, note down the load current and load voltage by changing the load in gradual steps. Maintain the speed of the generator constant during the test.
8. Care must be taken to not exceed the rated value of the generator/motor current.
9. Gradually remove the load and switch off the motor.
10. Draw the internal and external characteristics.

1.7.2.5 Observation table

Armature resistance R_a = _____ohms

Table 1.6: Observation Table.

S.No	Load voltage V_L (V)	Load current I_L (A)	Field current I_F (A)	Armature current $I_A = I_L + I_F$ (A)	$E_0 = V_L + I_A R_A$ (V)
1					
2					
3					

1.7.2.6 Model graphs

1. Generated voltage (E_g) versus load current (I_L)
2. Terminal voltage (V_t) versus load current (I_L)
3. Armature drop (I_aR_a) versus load current (I_L)

1.7.2.7 Practical calculations

The load test is conducted on shunt generator with the following rating and readings tabulated.

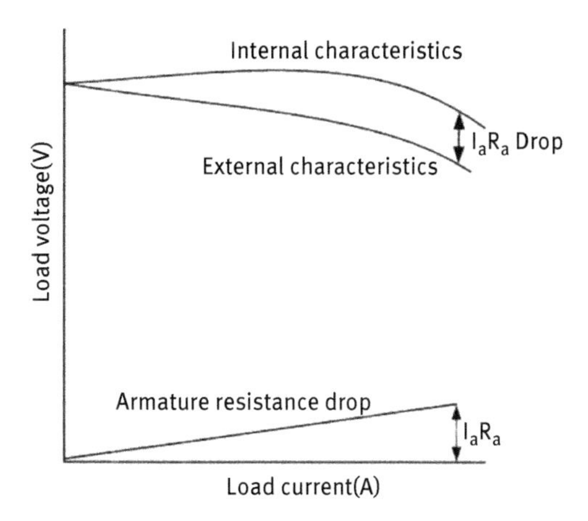

Fig. 1.19: Model graph of external characteristics of DC shunt generator.

Name plate details

Armature resistance R_a= 2.8 ohms

Table 1.7: Machine ratings.

Ratings	Motor	Generator
Rated Power	3.5 kW	3.5 kW
Rated Speed	1,500 rpm	1,500 rpm
Full Load Current	21 A	16 A
Excitation Current	0.5 A	0.9 A

1.7.2.8 Practical observations

Table 1.8: Practical observations.

S.No	Load voltage V_L (V)	Load current I_L (A)	Field current I_f (A)	Armature current $I_A = I_L + I_f$ (A)	$E_0 = V_L + I_a R_a$ (V)
1	220	0	0.76	0.76	222.12
2	210	1	0.74	1.74	214.87
3	200	2	0.72	2.72	207.61
4	190	3	0.7	3.7	200.36
5	180	4	0.68	4.68	193.10

1.7.2.9 Calculations

Generated emf E_g = Load Voltage + Armature Drop = $V_L + I_a R_a$

i) $E_g = V + I_a R_a = 220 + 0.76*2.8 = 222.12V$

ii) $E_g = 220 + 1.74*2.8 = 204.87V$

iii) $E_g = 180 + 2.72*2.8 = 187.61V$

iv) $E_g = 160 + 3.7*2.8 = 170.36V$

v) $E_g = 140 + 4.68*2.8 = 153.10V$

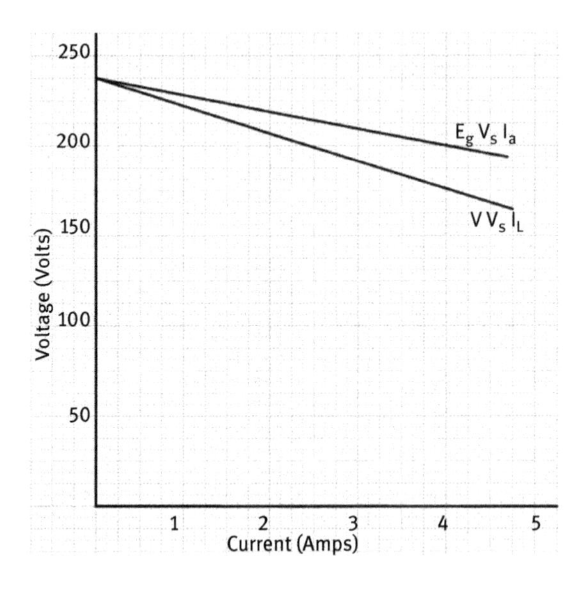

Fig. 1.20: DC Shunt generator external characteristics.

1.7.3 Testing of DC. series generator – load test

The external and internal characteristics DC series generator is drawn by conducting load test practically in laboratory by adopting the following method.

1.7.3.1 Apparatus required

Table 1.9: List of apparatus required.

Name of the equipment	Range	Quantity
Ammeter	0–20 A(MC)	1
Ammeter	0–5 A(MC)	1
Voltmeter	0–300 V(MC)	1

Table 1.9 (continued)

Name of the equipment	Range	Quantity
Voltmeter	0–3 V(MC)	1
Rheostat	400 Ω /1.7 A	1
Rheostat	100 Ω /5 A	1
Resistive Load	0–3,000 W	1
Tachometer	0–2,000 rpm	1

1.7.3.2 Theory

The field and armature windings in a series generator are connected in series, hence if the load is connected across the terminals the generator, the current drawn by the load is equal to the armature and field current. Therefore, the load, armature and field are in series.

Thus, $I_L = I_a = I_{se}$.

Performance characteristics

(i) No-load characteristic (magnetization characteristic) (E_g vs I_f)

These characteristics are obtained by drawing the curve between voltage generated (E_g) and field current (I_f). In series generator, to obtain this curve its field winding must be excited separately, i.e. the generator should be converted into separately-excited generator.

(ii) Internal characteristic (E_g vs I_a)

These characteristics are obtained by drawing the curve between voltage generated by the armature conductor (E_g) and armature current (I_a) at a specified speed of the generator.

External characteristics (V_L vs I_L)

These characteristics are obtained by drawing a curve between load voltage and current at a specified speed of the generator. The behavior of this characteristic is well understood by the following relation:

E_g = Voltage drop across the armature and field + Voltage across the external load,

$$E_g = I_a\,(R_a + R_{se}) + V_L \tag{1.21}$$

The nature of the curve is raising characteristic.

1.7.3.3 Circuit diagram

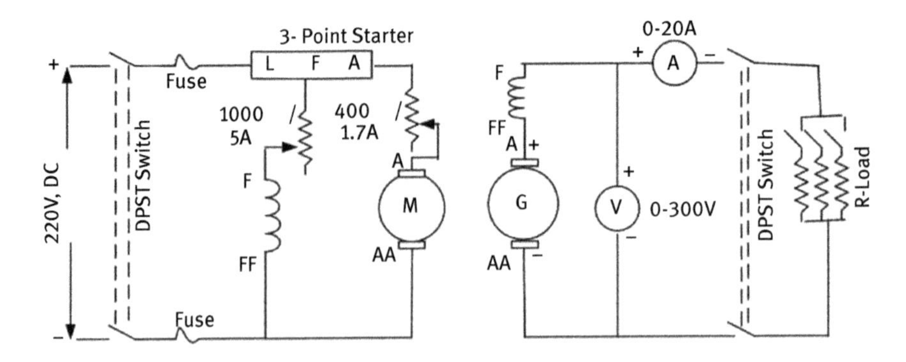

Fig. 1.21: Circuit diagram of DC series generator: load test.

Armature resistance (R_a)

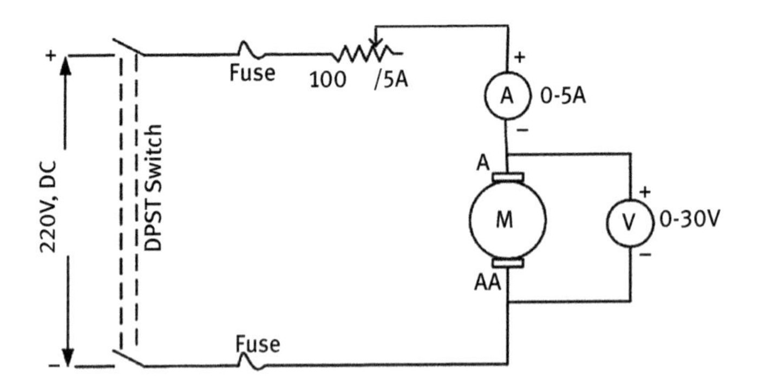

Fig. 1.22: Circuit diagram of DC series generator: armature resistance evaluation.

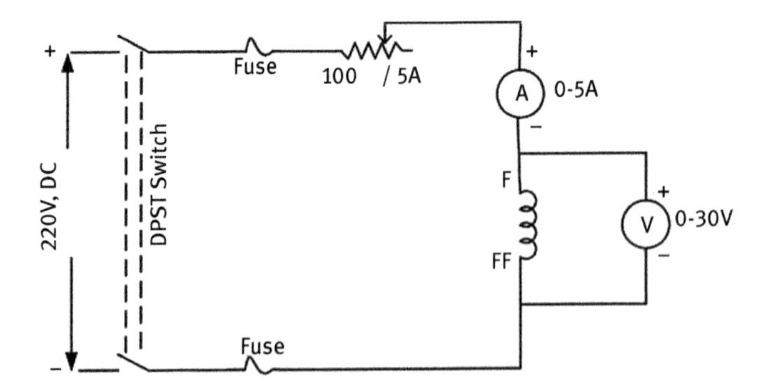

Fig. 1.23: Circuit diagram of DC series generator: field resistance evaluation.

Series Field resistance (R_{se})

1.7.3.4 Procedure

1. The name plate details of the shunt motor and generator should be noted.
2. The connections of various terminals are connected as per the circuit with DPST off position.
3. Switch on the supply (220 V DC) and using 3-point starter get the motor speed to its rated value by adjusting armature and field rheostat.
4. Now close the generator side DPST, note down the load current and load voltage by changing the load in gradual steps.
5. Maintain the speed of the generator constant during the test.
6. Care must be taken to not exceed the rated value of the generator/motor current.
7. Gradually remove the load and switch off the motor.
8. Draw the internal and external characteristics.

1.7.3.5 Observation table

Armature resistance of the generator =_____Ohms

Field resistance of the generator =_____Ohms

Rated speed of the generator =______Rpm

DC motor speed =______Rpm

Table 1.10: Observation table.

S.No	Armature voltage (V)	Load (kW)	$I_a (R_a + R_{se})$ (V)	Load current (A)	Generated voltage, E_g (V)
1.					
2.					
3.					

1.7.3.6 Model graphs

1. Armature voltage (V_a) versus armature current (I_a)
2. Terminal voltage (V_t) versus load current (I_L)

1.7.3.7 Practical calculations

The load is conducted on series generator with the following rating and readings tabulated:

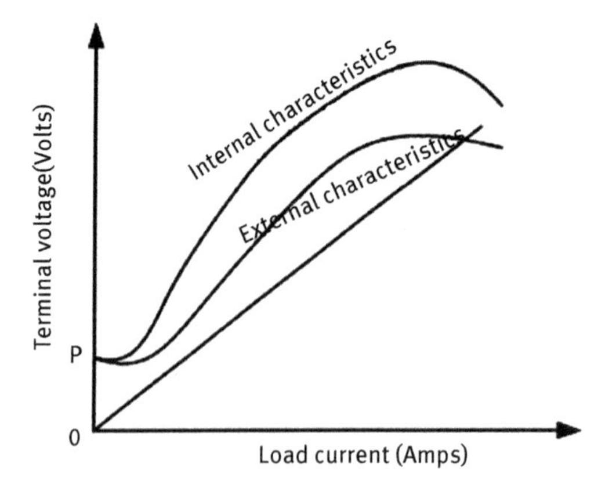

Fig. 1.24: Model graph of internal and external characteristics of DC series generator.

Power = 3.7 kW   Current = 19.5 A
Voltage = 220 V   Speed = 500 rpm

Table 1.11: Practical observations.

S.No	Armature voltage (V)	Load (kW)	I_a $(R_a + R_{se})$ Ω	Current flowing through load $I_a = I_L$ (A)	Voltage generated by the generator E_g (V)
1	100	3.58	22.62	8.7	122.62
2	119	2.75	16.38	6.3	135.38
3	90	2.5	13	5	103
4	45	2.25	5.72	2.2	50.72
5	40	2	4.68	1.8	44.68

I_L = Load current, I_a = Armature current.
R_a = Armature resistance = 1.5 Ω, R_{se} = Series resistance = 1.1 Ω
Generated emf $E_g = V + I_a\,(R_a + R_{se})$

i) $E_g = V + I_a\,(R_a + R_{se}) = 119 + 6.3^*(1.5 + 1.1) = 135.38\text{V}$
ii) $E_g = 90 + 5^*(1.5 + 1.1) = 103\text{V}$
iii) $E_g = 45 + 2.2^*(1.5 + 1.1) = 50.72\text{V}$
iv) $E_g = 40 + 1.8^*(1.5 + 1.1) = 44.68\text{V}$

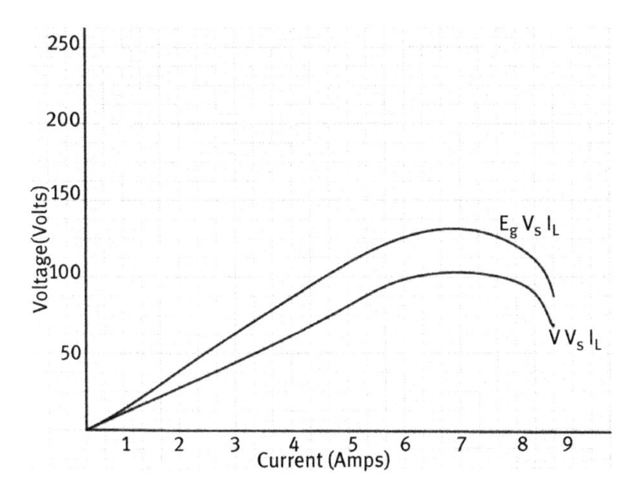

Fig. 1.25: Graph of internal characteristics of DC series generator.

1.7.4 Testing of D.C. compound generator - load test

The external and internal characteristics of cumulatively compound generator are drawn by conducting load test practically in laboratory by adopting the following method.

1.7.4.1 Apparatus required

Table 1.12: List of apparatus required.

Equipment	Range	Quantity
Voltmeter	0–300 V(MC)	1
Voltmeter	0–30 V(MC)	1
Ammeter	0–20 A(MC)	1
Ammeter	0–5 A(MC)	1
Ammeter	0–2 A(MC)	1
Rheostats	400 Ω/1.7 A	2
Rheostat	100Ω/5 A	1

Table 1.12 (continued)

Equipment	Range	Quantity
Resistive Load	0–3,000 W	1
Tachometer	0–2,000 rpm	1

1.7.4.2 Theory

A compound generator consists of series and shunt field winding. Based on the connection of series and shunt field winding it is classified into cumulatively and differentially compounded generator.

Cumulatively compound generator

In cumulatively compounded generator, the generated voltage increases with increasing the load current till the series field winding gets saturated. This is due to the combining of shunt and series field fluxes. Hence the total flux increases. After the saturation due to armature reaction, the terminal voltage decreases with increasing load current. Hence, theses generators are classified as flat, under and over compound generators based on the voltage generated.

Differentially compounded generator

In differentially compounded generator, the generated voltage decreases with increasing the load current. This is due to the fact that shunt and series field fluxes oppose each other. Hence the total flux decreases.

1.7.4.3 Circuit diagram

Cumulative compound generator

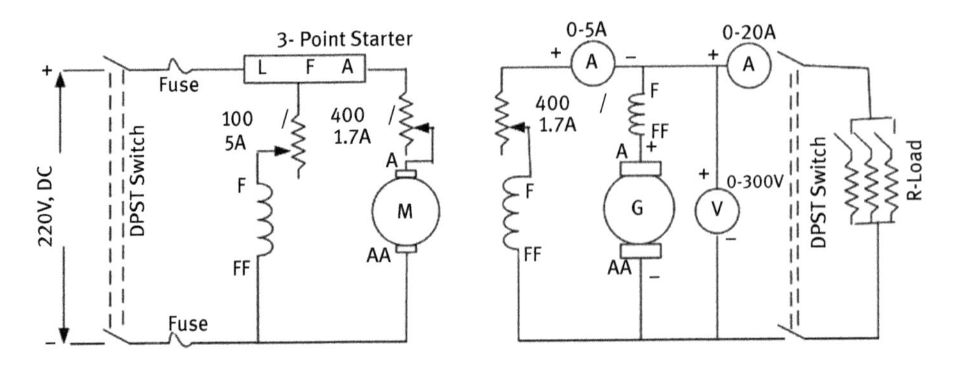

Fig. 1.26: Circuit diagram of DC cumulative compound generator: load test.

Differential compound generator

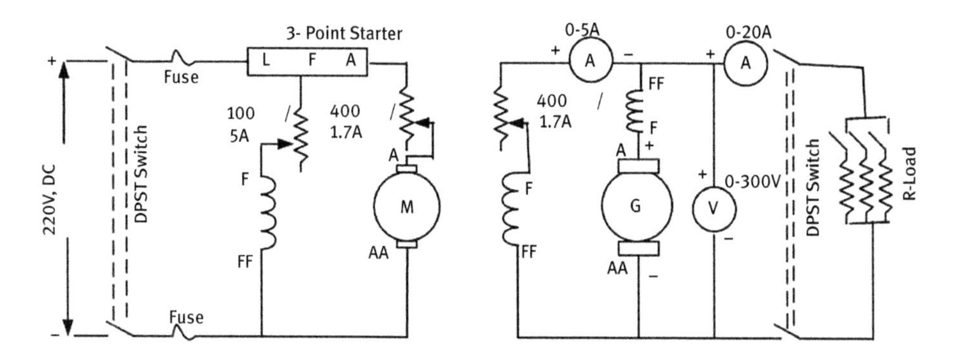

Fig. 1.27: Circuit diagram of DC differential compound generator: load test.

Armature resistance (R_a)

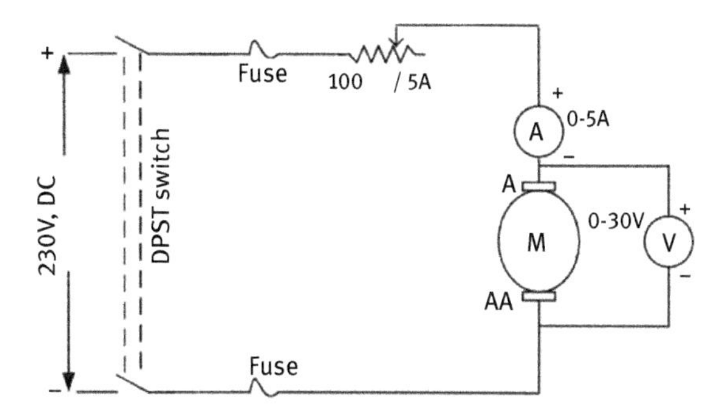

Fig. 1.28: Circuit diagram of DC compound generator: armature resistance evaluation.

Series field resistance (R_{se})

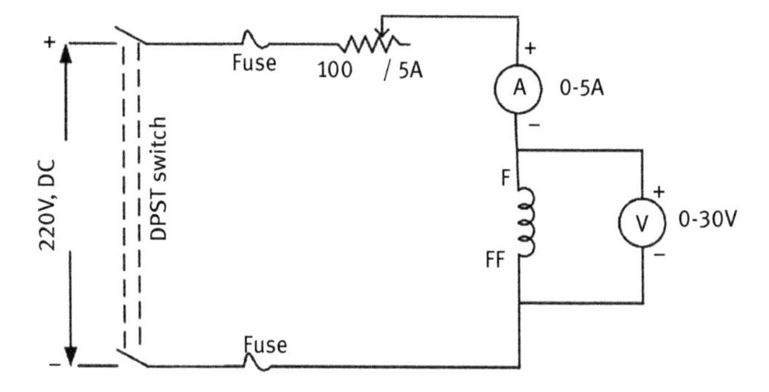

Fig. 1.29: Circuit diagram of DC compound generator: series field resistance evaluation.

1.7.4.4 Procedure

1. The name plate details of the shunt motor and generator should be noted.
2. The connections of various terminals are connected as per the circuit with DPST off position.
3. Switch on the supply (220 V DC) and using 3-point starter get the motor speed to its rated value by adjusting armature and field rheostat.
4. By adjusting field rheostat of the generator, bring the generator voltage at its rated value.
5. Now close the generator side DPST, note down the load current and load voltage by changing the load in gradual steps.
6. Maintain the speed of the generator constant during the test.
7. Care must be taken to not exceed the rated value of the generator/motor current.
8. Gradually remove the load and switch off the motor.
9. Draw the internal and external characteristics.

1.7.4.5 Observation table

Compound generator ratings

Motor	Generator
R_a =______Ohms	R_{sh} =______Ohms
R_f =______Ohms	R_{se} =______Ohms
	R_a =______Ohms
	N =______Ohms

Cumulatively compounded generator

Table 1.13: Observation table.

S.No	Load (kW)	Load current (amps)	Shunt field current (I_f)	Armature voltage (volts)	Armature current (amps)	I_a R_a ($I_a = I_f + I_L$)	Voltage generated $E_g = V + I_a R_a$ (V)
1							
2							

1.7.4.6 Model graphs

1. Terminal voltage (V_t) versus load current (I_L)
2. Generated voltage (E_g) versus armature current (I_a)

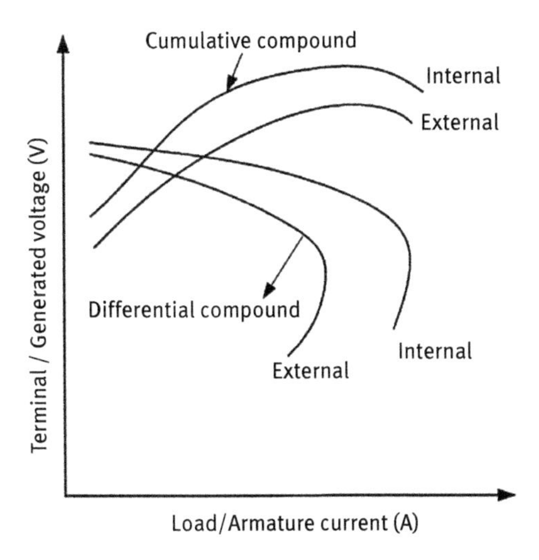

Fig. 1.30: Model graph of characteristics of DC cumulatively compounded generator.

1.7.4.7 Practical calculation

The load test on compound generator is conducted with the following rating and the readings tabulated:

Power = 3.7 kW	Current = 19.6 A
Speed = 1,500 rpm	Excitation = 0.56 A
Voltage = 220 V	

Table 1.14: Practical observations.

S.No	Load (kW)	Load current (A)	Shunt field current, I_f (A)	Armature voltage (V)	Armature current (A)	$I_a R_a$	Voltage generated $E_g = V + I_a R_a$ (V)
1	0	0	0.78	220	0.78	2.184	222.184
2	0.25	0	0.78	215	0.78	2.184	217.184
3	0.5	0.5	0.78	210	1.28	3.584	213.584
4	0.75	1	0.78	205	1.78	4.984	209.984
5	1	2	0.78	200	2.78	7.784	207.784

1.7.4.8 Calculations

R_a = Armature resistance = 2.8 Ω
I_L = Load current in amps
I_f or I_{sh} = Shunt field current in amps
I_a = Armature current in amps = $I_L + I_F$
E_g = Generated emf in Volts

i) $E_g = V + I_a R_a = 220 + 0.78*2.8 = 222.184V$
ii) $E_g = 215 + 0.78*2.8 = 217.184V$
iii) $E_g = 210 + 1.28*2.8 = 213.584V$

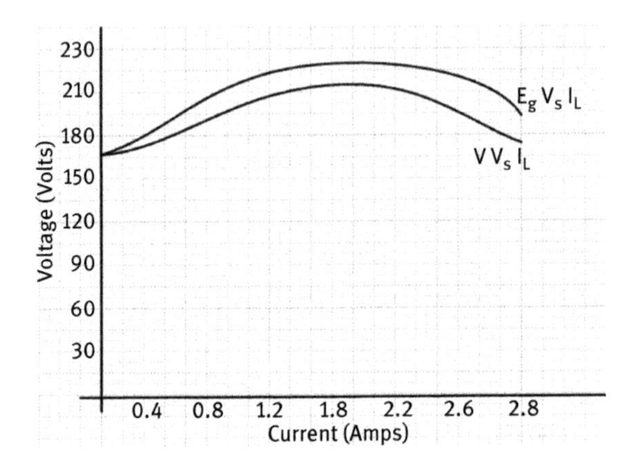

Fig. 1.31: Graph of external characteristics of DC cumulatively compounded generator.

1.7.4.9 Result

The internal and external characteristics of cumulatively compound generator are drawn by conducting load test.

1.8 Problems

1. The shunt generator running at 1,000 rpm and the voltmeter and ammeter readings are

Ammeter I_f (A)	2	4	6	8	10
Voltmeter E_g (V)	162	262	394	469	524

Find out:

i) The open circuit voltage required to excite.
ii) Critical field resistance.

Assume $R_a = 0.42\ \Omega$ and $R_{sh} = 70\ \Omega$

Solution:

(i) Plot the magnetization curve as shown below and draw the field resistance line (70 Ω). The meeting point of shunt resistance line and OCC gives the open circuit voltage of 540 V.

(ii) To find the critical field resistance, calculate the slope of the straight part of the OCC

$$\text{Critical field resistance} = \frac{\text{Opencircuit Voltage}}{\text{Field Current}} = \frac{150}{1} = 150\ \Omega$$

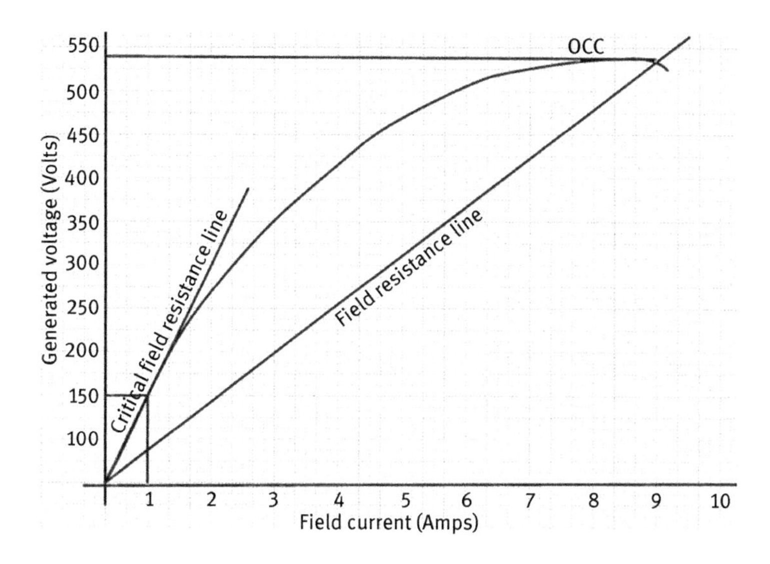

2. The load test is conducted on series motor and the recorded data are as follows:

Current:	20	30	40	50
Torque:	128.8	230.5	349.8	46.2

Draw the speed–torque curve for the machine when supplied at a constant voltage of 460 V. Resistance of the armature resistance is 0.5 Ω.

Solution:

For current $I_a = 20$ A
Motor input power = 460*20 = 9,200 W
Field and armature Cu losses = $I_a^2 * R_a = 20^2 * 0.5 = 200$ W
Motor output power = Motor input – Losses = 9,200 – 200 = 9,000 W
Output power = 2 πNT/60
N = 60 * Output power/2 πT = 9,000/2 * 3.14* 128.8 = 667 rpm

Similarly, for all the values of current, the speed is calculated and tabulated.

Current (A)	**20.2**	**30.5**	**40.0**	**49.9**
Torque (N–m)	128.8	230.5	349.8	46.2
Losses (Watts)	200	450	800	1,250
Output (Watts)	9,200	13,350	17,600	21,850
Speed (rpm)	667	551	480	445

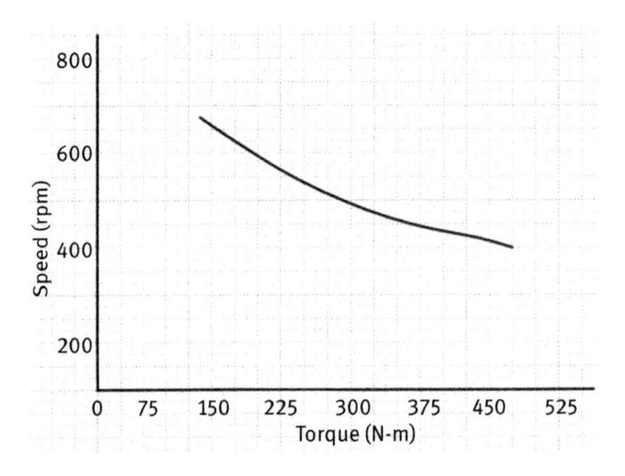

1.9 Viva-voce questions

Magnetization characteristics

1. The magnetization curve differs for increasing and decreasing values of field current. Why?
2. What is the function of armature conductors in generator?
3. The speed is kept constant during the no-load test of the generator. Why?
4. What is function of pole shoes in a generator?

5. Define the residual magnetism.
6. What is meant by critical speed and critical field resistance?
7. What is procedure to find critical field resistance using no-load curve?
8. Mention the uses of magnetization curve.
9. What is cross magnetizing effect?
10. What is demagnetizing effect?

Load test on DC shunt generator

11. Mention the reasons for why no emf is built up by the generator.
12. Define armature reaction.
13. Why the shunt generator characteristics have drooping nature?
14. The DC generators are designed for highest efficiency. Why?
15. If a capacitive load is connected across a generator, what will happen?
16. The DC generators overall efficiency in the order of ——%.
17. Define efficiency of DC generators.
18. Mention types of losses in a DC generator.
19. Plot the shunt generator external and internal characteristics.
20. Write the formulas for eddy current and hysteresis losses in generator.

Load test on series generator

21. Plot the external and internal characteristics of a series generator.
22. Mention the causes for the failure of voltage build-up by the series generator.
23. What are the applications of DC series generator?
24. While running the series generator at its rated speed, what happens if the field winding is opened suddenly?
25. The substances used for the brushes are ________ Why?
26. Why silica is added to the steel.
27. Define permeability.
28. How do you convert shunt generator into series generator?
29. Write voltage equation generator.
30. While running the series generator at its rated speed, what happens if the load is disconnected suddenly?

Load test on compound generator

31. How many windings were present in compound generator?
32. Name the two windings of a compound generator.
33. Define level compound generator.
34. Which field winding dominates in a compound wound generator.
35. While running the compound generator at its rated speed, what happens if the shunt field is disconnected suddenly?
36. What is the function of commutator in a DC generator?

37. While running the compound generator at its rated speed, what happens if the series field is disconnected suddenly?
38. Define flat compound generator.
39. What is the difference between cumulative compound and differential compound generators?
40. Write the applications of compound generators.

1.10 Objective questions

1. The ——— generator produces voltage without residual magnetism.
   (a) Compound (b) Series (c) Shunt (d) Self-excited

2. The resistance of— generator is called critical resistance.
   (a) Field (b) Brushes (c) Armature (d) Load

3. In a generator, when copper losses = constant losses, then the efficiency is will be
   (a) zero (b) minimum (c) medium (d) maximum

4. The residual magnetism present in a DC machines is in the order of—-
   (a) 2 to 3% (b) 1 to 2% (c) 2 to 5% (d) 5 to 7%

5. The mmf is large in –
   (a) air gap (b) coil (c) inductance (d) core

6. For generating low voltage and high current — -lap winding is used.
   (a) wave (b) lap (c) Either wave or lap (d) Both a & b

7. In DC generator, the magnetic flux is created by
   (a) commutator (b) permanent magnets (c) brushes (d) None

8. The ——— generator produces voltage with residual magnetism.
   (a) series (b) shunt (c) compound (d) All

9. The voltage induced at zero speed of the generator due to the residual magnetism is
   (a) less (b) equal the rated speed (c) zero (d) high

10. A shunt generator running at 1,000 rpm and 200 V is generated. If the generator speed is 1,200 rpm the generated voltage will be
    (a) 200 V (b) 100 V (c) 240 V (d) 300 V

11. The use of dummy coils in a generator is to
    (a) decrease hysteresis losses
    (b) improve flux
    (c) increase the voltage

(d) provides the rotor balance mechanically

12. In a shunt generator, the voltage generated in the armature is 600 V; the resistance of armature is 0.1 Ω. If the current flowing through armature conductor is 200 A, the terminal voltage will be
(a) 640 V (b) 620 V (c) 600 V (d) 580 V

13. If mechanical load decreases then the speed of the DC shunt motors
(a) increase abruptly
(b) does not depends on the load
(c) will remain almost constant
(d) will decrease

14. If AC supply is applied to DC motor, it will
(a) run as induction motor
(b) run as synchronous motor
(c) not run
(d) burn

15. The effect sparking in a DC motor may
(a) harm the segments of the commutator
(b) harm the insulation of commutator
(c) Increase power consumption
(d) All of the above

16. The power developed by the motor is maximum when the ratio of E_b/V is equal to
(a) 4.0 (b) 2.0 (c) 1.0 (d) 0.5

17. The relative motion between the flux and conductor in a generator to induce the voltage
(a) should be parallel
(b) should not be parallel
(c) Both (a) & (b)
(d) None

18. A 220 V, 190 A is delivered by the shunt generator. The copper and stray loss are 2,000 W and 1,000 W respectively. Then the generator efficiency is
(a) 81.65% (b) 82.70% (c) 93.30% (d) 99.13%

19. A 120 V, 1,200 rpm DC shunt motor is operated on unloaded. An extra resistance of 4 Ω is added in series with the shunt field then the speed increases to 1,370 rpm by keeping the same terminal voltage. Then the value of series resistance is
(a) 28.23 Ω (b) 44.44 Ω (c) 33.33 Ω (d) 66.66 Ω

20. In a series motor, the torque is proportional to
    (a) V_a (b) I_a^2 (c) $1/I_a^3$ (d) $1/V_a$

21. If the current drawn by the motor is increased from 15 amps to 18 amps, then the % of the torque increased as a percentage of initial torque is
    (a) 11% (b) 22.22% (c) 44% (d) 33.33%

22. In a wave wound series motor P = 4 V = 240 V, Z = 180 conductors. The R_a and R_f are 0.10 Ω and 0.2 Ω. If the motor is taking 40 A current at 0.015 Wb flux per pole, then the speed of the motor is
    (a) 2,533.3 rpm (b) 1,664.6 rpm (c) 1,256.8 rpm (d) 1,897.7 rpm

23. In belt drive applications DC series motor is used Y/N

24. For —, differentially compound motors is used
    (a) high running torque
    (b) low starting torque
    (c) fixed speed
    (d) variable speed

25. — motor is used in elevators
    (a) Induction motor
    (b) Synchronous motor
    (c) Differential compound motor
    (d) Cumulative compound motor

Answers

1. d, 2. a, 3. d, 4. a, 5. a, 6. b, 7. b, 8. d, 9. c, 10. c, 11. d, 12. d, 13. c, 14. d, 15. d, 16. d, 17. b, 18. c, 19. a, 20. b, 21. c, 22. a, 23. N, 24. b, 25. d.

1.11 Exercise problems

1. In DC machine, the total iron loss is 10 kw at its rated speed and excitation. If excitation remains the same, but speed is decreased by 25%, the total iron loss is found to be 5 kw. Calculate the hysteresis and eddy current losses at (i) full speed (ii) half the rated speed.
2. A 10 kW, 250 V DC shunt generator has a total load rotational loss of 400 watts. The armature circuit and shunt field resistances are 0.5 ohm and 250 ohm respectively. Calculate the shaft power input and the efficiency at rated load. Also calculate the maximum efficiency and the corresponding power output.
3. The data is recorded by conducting Hopkinson's test.
   Voltmeter reading connected across the supply = 230 V
   Motor input current drawn from the mains = 2 A

Current drawn by the motor from the generator = 5 A
Motor field current = 1.7 A
Generator field current = 2.3 A
The armature resistance of each machine = 0.02 Ω
Calculate efficiency of both the machines.

2 DC Motors

2.1 Introduction

A machine that transforms electric energy into mechanical energy is known as motor. The major applications of DC motor are buses, airplanes and cars, injection pump and cooling fan. Whenever a conductor carries current and is kept in a magnetic field, it experiences a force according to Lorenz's law. The direction of the force experienced by the conductor is determined by the Fleming's left-hand rule.

2.2 Principle operation of DC motor

A conductor that is free to rotate in a magnetic field is shown in Fig. 2.1. A DC current flows through the rectangular conductor that is supplied by the brushes situated on a commutator.

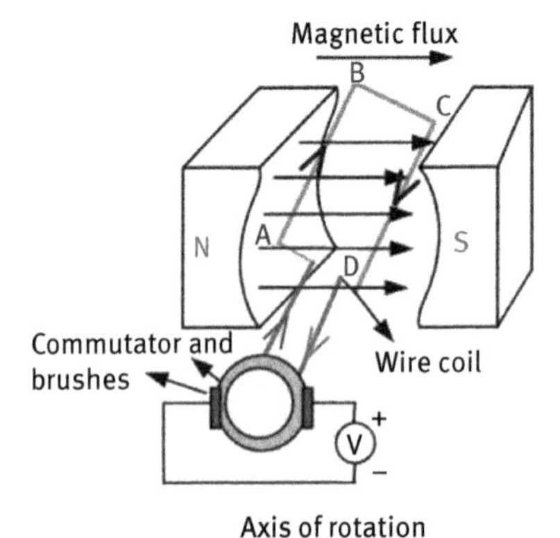

Fig. 2.1: DC motor principle.

If a current-carrying conductor is placed in magnetic field, the magnetic field setup in the conductor and the field setup by the permanent magnets interact and a force F is exerted on the conductor. This force causes torque and the coil rotates. Thus, electrical energy is converted into mechanical energy.

2.3 Types of DC motors

Based on the excitation of field windings, DC motors are categorized as separately and self-excited. Further, the self-excited motors are categorized as (a) shunt, (b) series and (c) compound.

https://doi.org/10.1515/9783110682274-002

2.4 EMF equation of a motor

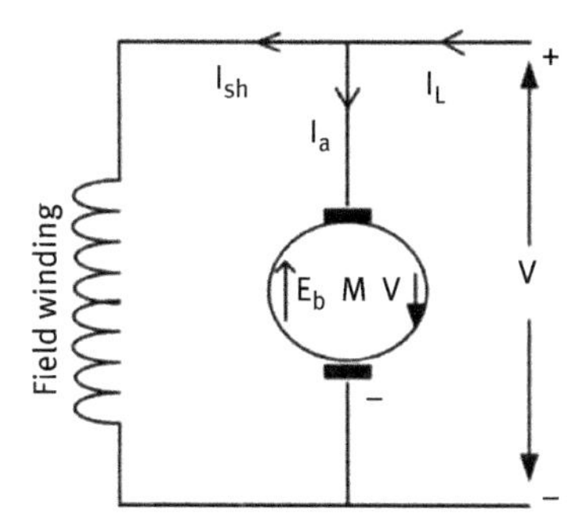

Fig. 2.2: Equivalent circuit of DC motor.

The supply voltage connected across terminals of the motor as shown in Fig. 2.2. will be utilized to conquer the back emf (E_b) and voltage drop due to $I_a R_a$.

Therefore,

$$V = E_b + I_a R_a \tag{2.1}$$

The mechanical power output P_m = input power − losses

$$P_m = V^* I_a - I_a^{2*} R_a \tag{2.2}$$

The condition for maximum power is $\frac{d}{dI_a}(P_m) = 0$.

We get,

$$I_a R_a = V/2 \quad \text{(substituting in (2.1))}$$

$$V = E_b + \frac{V}{2}$$

$$E_b = \frac{V}{2} \tag{2.3}$$

Thus, the power developed by the motor is maximum when the back emf (E_b) is equal to half the supply voltage (V/2).

a) Shunt motor

The DC shunt motor is shown in Fig. 2.3, it has shunt field winding and armature winding and are connected in parallel. The shunt field carries very low current due to high resistance. Since the field current has low effect on the strength of the magnetic field, the motor speed is not affected significantly by disparity in load current:

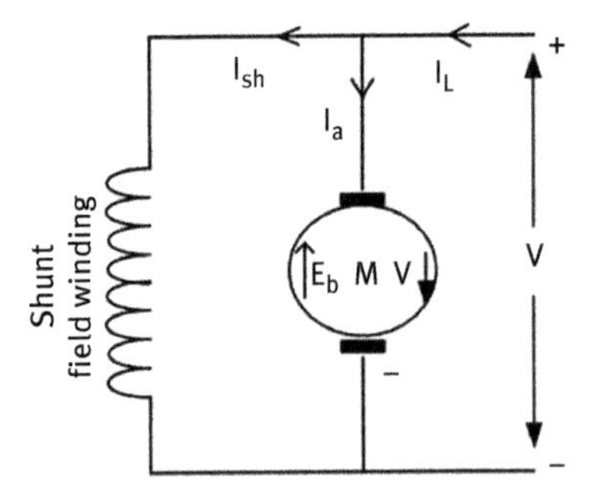

Fig. 2.3: Equivalent circuit of DC shunt motor.

$$V = E_b + I_a R_a \tag{2.4}$$

$$I_a = I_L + I_f \tag{2.5}$$

b) Series motor

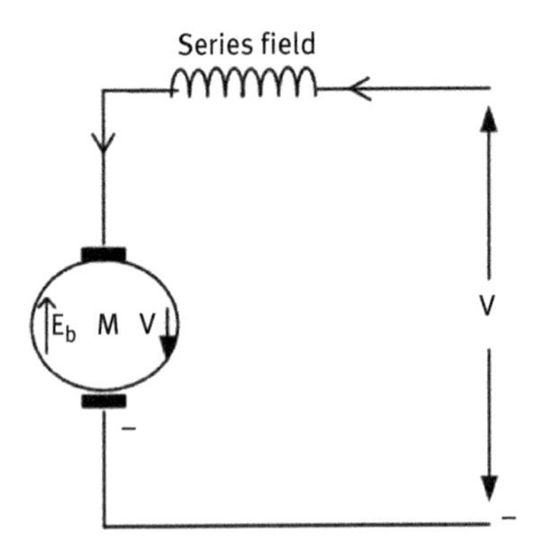

Fig. 2.4: Equivalent circuit of DC series motor.

$$V = E_b + I_a R_a \tag{2.6}$$

$$I_a = I_L = I_{se} \tag{2.7}$$

In series motor, the field winding and armature winding are connected in series as shown in Fig. 2.4:

$$I_a = I_L = I_f$$

The field winding consists of less turns having larger diameter wire and it has low resistance. Any change in load on the shaft causes the variation in current through the field. If the load increases, the armature current also increases and produces a strong magnetic field.

c) Compound motor

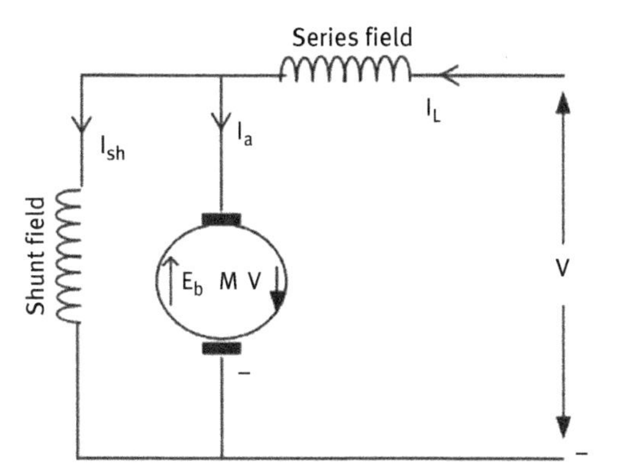

Fig. 2.5: Equivalent circuit of DC compound motor.

The compound motor consists of two field windings: one in series and one in parallel with the armature as shown in Fig. 2.5. Therefore, it exhibits series and shunt motor characteristics. It has high torque similar to series motor and good speed regulation similar to shunt motor.

There are two types of compound motors, namely, cumulative and differential compound motor.

2.5 Torque

The rotating or twisting movement of an object is called torque. It can be also defined as the product of force and perpendicular distance:

$$T = F \times R \tag{2.8}$$

Consider a pulley as shown in Fig. 2.6 of radius “r” meters acted upon by a circumferential force, F which causes it to rotate at N rpm.

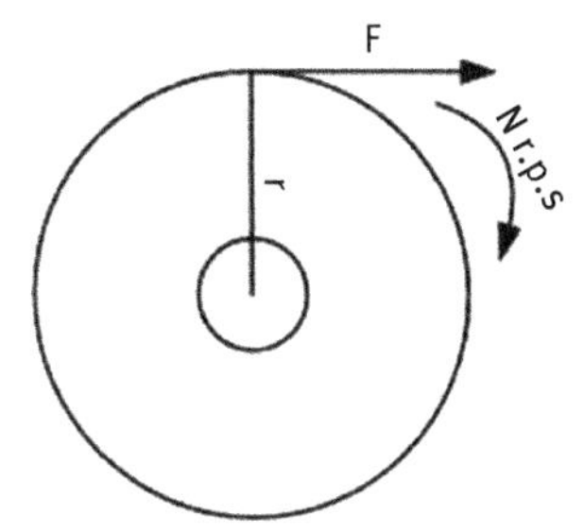

Fig. 2.6: Pulley.

$$\text{The torque } T = F \times r \text{ N–m.}$$

$$\text{Work done by the force in one revolution} = \text{Force} \times 2\,\pi r$$

$$\text{Power developed}(P) = T \times 2\pi r \times N \text{ J/s or W}$$

$$= T \times \omega = \frac{2\pi NT}{60} \quad (\text{N is in RPM}) \qquad (2.9)$$

2.6 Characteristics of DC motor

The three important characteristics for DC motors considered are
(i) Torque (T) versus armature current (I_a)
(ii) Speed (N) versus armature current (I_a)
(iii) Speed (N) versus torque (T)

The above relations are explained by the following relations:
Armature torque (T_a) $\propto \phi . I_a$ and
Speed (N) $\propto E_b/\phi$

For a DC motor, the back emf E_b is given by

$$\text{i.e. } E_b = \phi\, ZNP/60\, A.$$

Z, P and A are constants for any machine.
Therefore, $N \propto Eb/\phi$

a) Characteristics of DC shunt motors

Torque (T) versus armature current (I_a)

The torque developed by the shunt motor (T_a) $\propto \phi . I_a$

For shunt motor, the shunt field flux is constant. Therefore, the torque is proportional to armature current. Hence, the characteristics will be a straight line through the origin shown in Fig. 2.7. The shunt motors are never started with heavy loads because motor requires high current.

Speed (N) versus armature current (I_a)

As flux ϕ is constant, the speed N is proportional to back emf E_b. As back emf is constant, the speed remains constant. In general, flux ϕ and back emf E_b reduce with increase in load. The back emf E_b decreases a little higher than flux ϕ and the speed reduces slightly. Therefore, a shunt motor is assumed as a constant speed motor. The characteristics are shown in Fig. 2.8. The straight horizontal line indicates the ideal characteristic and the actual characteristic is represented by the dotted line. Practically, 5–10% of the speed is decreased.

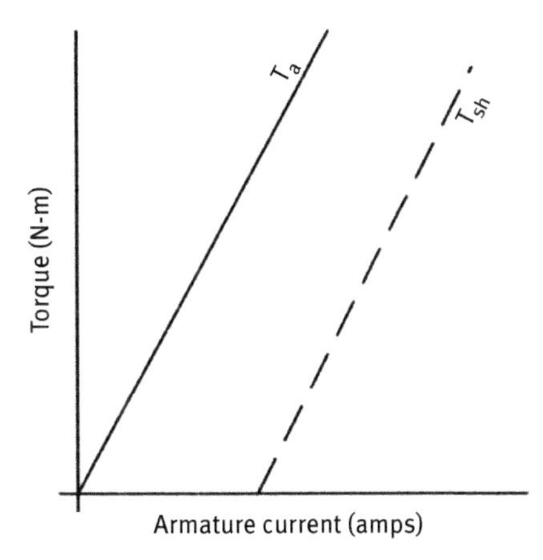

Fig. 2.7: Torque versus armature current.

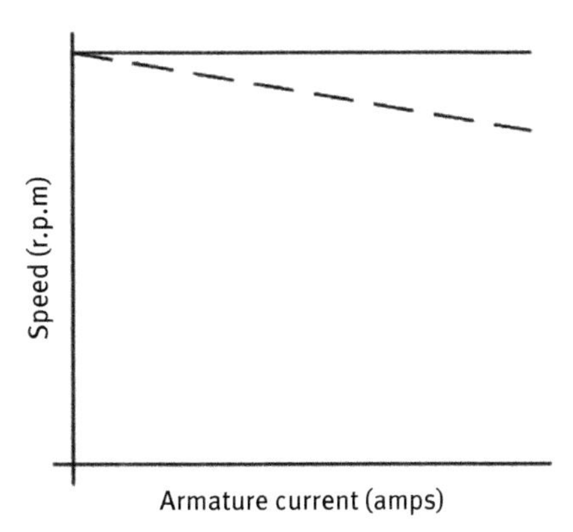

Fig. 2.8: Speed versus armature current.

Speed (N) versus torque (T)

The relation between speed and torque is essential in deciding a DC motor for exact purpose.

$$\text{Power developed by a DC motor} = \frac{2\pi NT}{60}.$$

From the above equation $N \propto \frac{1}{T}$. The speed of the motor is inversely proportional to the torque as shown in Fig. 2.9.

b) Characteristics of DC series motors

Torque (T) versus armature current (I_a)

We know that $T_a \propto \phi.I_a$. In a series motor, $\phi \propto I_a$. Hence, the torque developed by the series motor up to the saturation of magnetic field $T_a \propto I_a^2$. Therefore, the torque versus armature current (T_a–I_a) plot is a parabola for lesser values of armature current (I_a) as shown in Fig. 2.10.

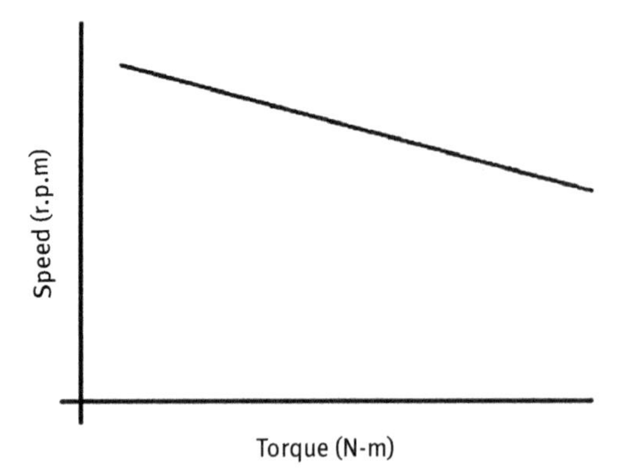

Fig. 2.9: Speed versus torque.

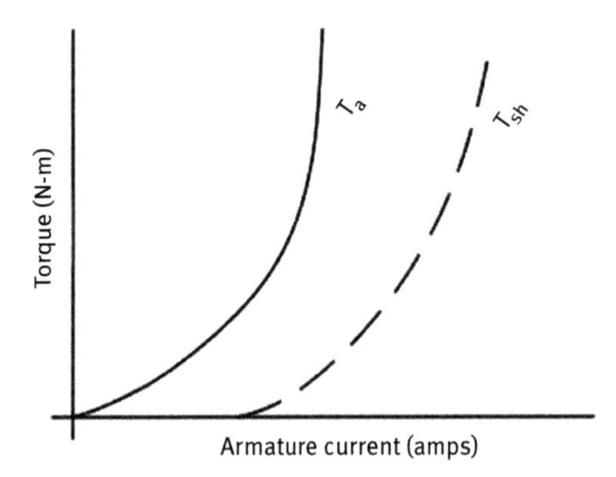

Fig. 2.10: Torque versus armature current.

The magnetic field gets saturated when the field current reaches the saturated value of the field, then magnetic field flux ϕ does not depend on the armature current I_a. Therefore, the torque varies proportionally to I_a only, $T \propto I_a$. Therefore, above the magnetic field saturation, T_a–I_a graph becomes a straight line. The shaft torque (T_{sh}) is less than the armature torque (T_a) due to stray losses. Hence, the graph T_{sh} versus I_a presents somewhat below the T_a.

Speed (N) versus armature current (I_a)

We know that the speed (N) of a DC motor is proportional to E_b/ϕ.

In series motor $I_a = I_f = I_L$, and the flux developed by the motor is proportional to the field current. Therefore, the speed of the series motor is inversely proportional to armature current, that is, $N \propto \frac{1}{I_a}$. When the load current is small the field current is small. Hence, the motor runs at very high speeds and as the load current increases the speed decreases. Therefore, when the armature current is very less, the speed becomes seriously high as shown in Fig. 2.11. Due to this, series motor is never started without some load.

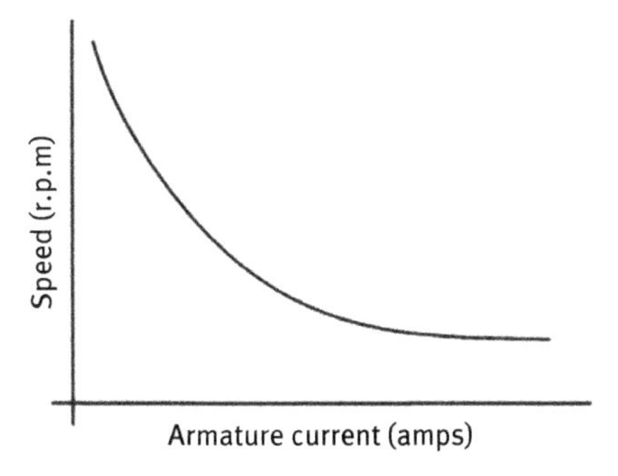

Fig. 2.11: Speed versus armature current.

Speed (N) versus torque (T_a)

$N \propto \frac{1}{I_a}$ and $T_a \propto {I_a}^2$; hence, $N \propto \frac{1}{\sqrt{T}}$. From the relations it shows that as the torque of the increases speed decreases and vice versa as shown in Fig. 2.12.

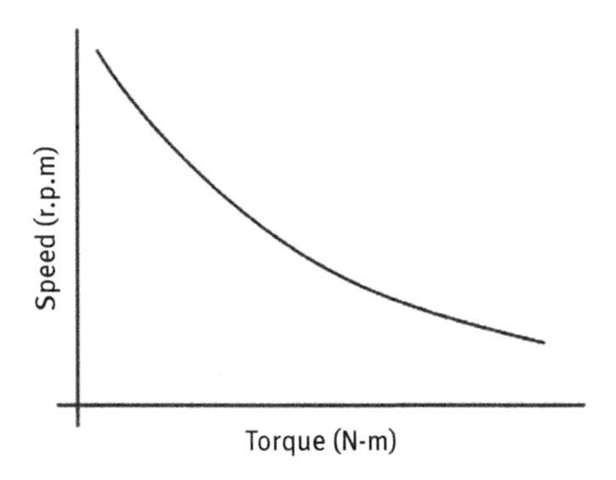

Fig. 2.12: Speed versus torque.

c) Characteristics of DC compound motor

(i) Cumulative compound motor

These motors are employed where series characteristics are necessary. Series field winding helps in carrying the heavy load currents and the shunt field winding stops the motor from running at high speed when the load is removed suddenly.

(ii) Differential compound motor

In this motor, the series field flux opposes shunt field flux. Therefore, the total flux reduces with an increase in load as shown in Fig. 2.13. Because of this, the speed remains almost steady or even it may increase a little with an increase in load $(N \propto E_b/\phi)$.

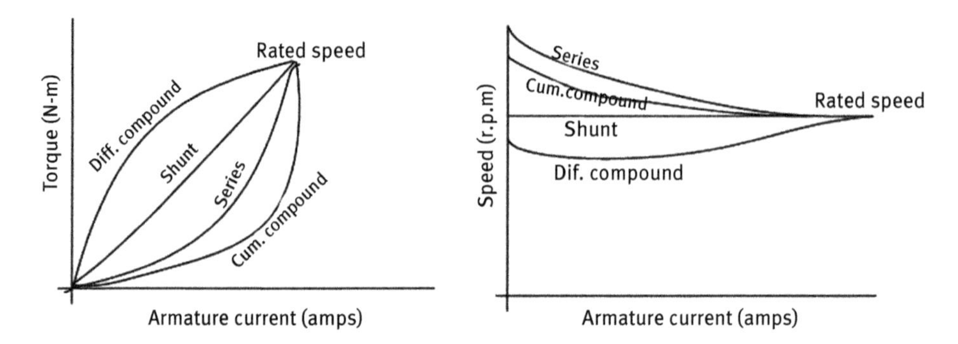

Fig. 2.13: Characteristics of DC compound motor.

2.7 Losses

While transforming electrical power into mechanical power, the entire electrical power is not converted into mechanical output. Hence, a little portion of the input power wasted in various parts of the machine is known as loss of power or loss. The major drawbacks of these losses are rise in temperature, reduction in the efficiency and reduction in the lifetime of the machine.

These losses are of two types: (i) constant losses and (ii) variable losses.

(i) Constant losses

The constant losses can be further divided as eddy current loss and hysteresis loss

a) Hysteresis loss

The magnetic flux developed in the armature winding undergoes reversal of magnetic field in one revolution, as the rotation of armature conductors is a continuous process. Hence, the reversal of magnetic field is also continuous. This magnetic field links with stator and rotor core of the machine, but some part of the energy lost in the core due to reversal of the magnetic field and is called hysteresis loss. This loss is constant and it is minimized by making the core with high grade of material.

The hysteresis of loss is calculated using the Steinmetz formula.

$$\text{Hysteresis loss, } P_h = \eta\, B_{max}^{1.6}\, f\, V \tag{2.10}$$

where η is the Steinmetz coefficient of hysteresis;
B_{max} is the maximum flux density in the armature winding;
f is the magnetic reversals frequency;
V is the armature core volume in m^3.

b) Eddy current loss

If the magnetic flux developed by the field links with armature conductors, an emf is produced in the conductors according to the Faradays laws of electromagnetic induction. The voltage also induced in the armature core because armature conductors are mounted on the core. Therefore, a small magnitude of current flows in the core, which is not useful and this current is called eddy current. The loss due to the eddy current is called eddy current loss. This eddy current loss is constant and it is minimized by making the core with thin laminations.

$$\text{Eddy current loss}, P_e = \eta B_{max}^2 f^2 t^2$$

where t is the thickness of the laminations.

ii) Variable losses

The variable loss is also called as copper loss (I^2R) and is further divided into (a) armature copper loss; (b) field copper loss and (c) brush contact loss.

a) Armature copper loss

If the current flowing the armature having resistance "R_a" is "I_a" amperes, then the loss due to this current is $I_a^2R_a$. This loss depends on the armature current or load current that is variable. Hence, these losses are termed as variable losses or copper losses. These losses contribute to 30% of the total losses:

$$\text{Armature copper loss} = I_a^2 R_a \tag{2.11}$$

Thus, the armature copper losses are proportional to the square of the current.

b) Field copper loss

If the current flowing the field winding having resistance "R_f" is "I_f" amperes, then the loss due to this current is $I_f^2R_f$. This loss depends on the field current which is constant. Hence, these losses are also treated as constant losses or field copper losses. These losses contribute to 25% of the total losses.

$$\text{The field winding copper loss} = I_f^2 R_f \tag{2.12}$$

c) Brush contact loss

The losses occur due to the brush contact on the commutator. It is usually 2–4 V for carbon brushes.

Stray load loss

There are some losses other than the losses that have been explained above are called stray losses. These losses are due to the commutation, armature reaction and other losses that are not easy to find. These losses constitute 1–2% of the total losses.

2.8 Efficiency

The performance of the motor is measured in terms of efficiency and all the motors are designed to exhibit good efficiency. The efficiency of a motor is defined as the ratio between output power to input power. The input power and output power are calculated using the below equations. The power flow diagram of the motor is shown in Fig. 2.14.

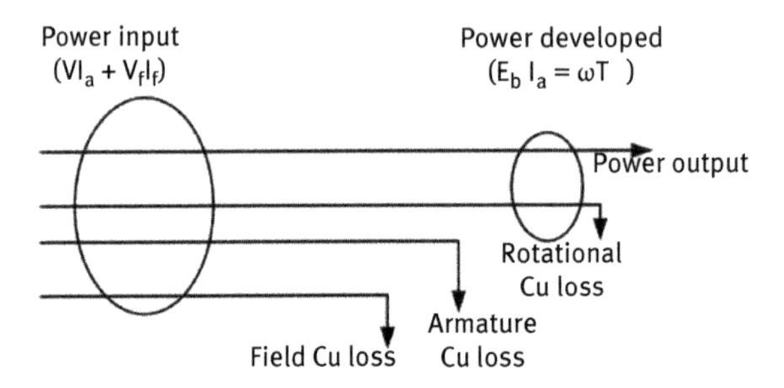

Fig. 2.14: Power flow in a DC motor.

$$\text{Power output} = \text{power developed} - \text{rotational losses.}$$

The efficiency of the DC motor can be calculated as follows:

$$\text{Efficiency}\ \eta = \frac{P_{out}}{P_{in}} \times 100 \tag{2.13}$$

This equation can also be expressed in terms of power losses in the motor:

$$\text{Efficiency}\ \eta = \frac{P_{out}}{P_{out} + P_{cu-loss} + P_{field-loss} + \text{rotational losses}} \times 100 \tag{2.14}$$

Condition for maximum efficiency

$$\frac{d(\text{efficiency})}{dt} = 0 = \frac{d}{dt}\left(\frac{VI}{VI + I^2 R + P_c}\right) = 0 = I^2 R = P_c \tag{2.15}$$

Hence, the efficiency is maximum when variable losses = constant losses.

Thus, the efficiency increases with an increase in load current reaches a maximum value when load current equals to $I = \sqrt{\frac{Pc}{R}}$ and then starts decreasing. The motor is designed to give maximum efficiency at the rated output of the machine. A typical efficiency curve is shown in Fig. 2.15.

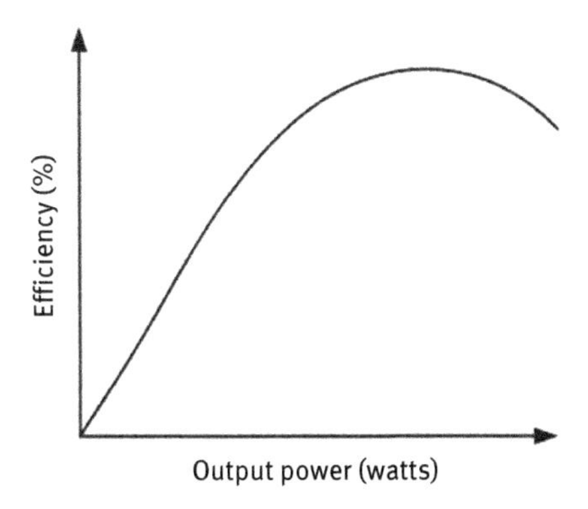

Fig. 2.15: Efficiency versus output power.

2.9 Testing of DC motors

To assess the performance of the machines, such as efficiency, machine characteristics and losses have to be verified or tested after DC machines are designed. From the tests, the characteristics of both internal and external losses and efficiencies are calculated.

2.9.1 Testing on DC shunt motor-brake test

The load test or brake test is conducted on DC shunt motor to obtain the performance curves.

2.9.1.1 Apparatus required

Table 2.1: List of required apparatus.

Equipment	Range	Quantity
Ammeter	0–20 A (MC)	1
Ammeter	0–2 A (MC)	1
Voltmeter	0–300 V (MC)	1
Rheostat	400 Ω /1.7 A	1
Rheostat	100 Ω /5 A	1
Tachometer	0–2,000 rpm	1

2.9.1.2 Theory

Brake test is a direct technique of testing the DC machine in which output power, torque, efficiency and speed are determined at different load conditions.

The back emf, $E_b = V - I_a R_a$ and $E_b = K_\phi N$:

$$\therefore K_\phi = E_b/N \tag{2.16}$$

where V is the supply voltage,
I_a is the current flowing through armature conductors,
R_a is the resistance of the armature conductors.

Power drawn by the motor (P_{in}) = power drawn by the field + power drawn by the armature:

$$P_{in} = V_f I_f + V_a I_a \tag{2.17}$$

If the drum radius is "r," torque developed in the drum is calculated by the equation

$$T_{shaft} = 9.81(T1 \sim T2) \tag{2.18}$$

$$\text{Output power developed by the motor } (P_{out}) = T_{shaft} \times \omega$$
$$= 9.81 * r * (T1 \sim T2)\ 2\pi N/60 \tag{2.19}$$

$$\%\,\text{Efficiency} = \frac{P_{out}}{P_{in}} \times 100 \tag{2.20}$$

The voltage equation of the motor is given by

$$V = E_b + I_a R_a \tag{2.21}$$

Based on eq. (2.16), when $E_b = 0$, the speed increases and the back emf also increases gradually.

In a motor the electrical power supplied = mechanical power developed by the rotor

$$= E_b I_a = T\omega \tag{2.22}$$

where $\omega = 2\pi N/60$ is the angular speed of the drum, in rad/s.

From eq. (2.22), the torque developed by the motor is given by the relation

$$T \propto \phi I_a \tag{2.23}$$

In a shunt motor, the shunt-filed flux is constant.

Torque (T) versus armature current (I_a) characteristics

From eq. (2.23), it shows that the torque developed by the motor is proportional to the armature current when flux ϕ is constant.

Torque (T) versus speed (N)

The torque requirement increases as the load on the motor increases. To meet this, motor draws additional current from the supply as this armature voltage drop increases slightly. Hence, the speed also drops slightly since shunt field flux is constant.

Output power (P_o) versus efficiency (η)

The efficiency of the motor increases with the increased output power.

2.9.1.3 Circuit diagram

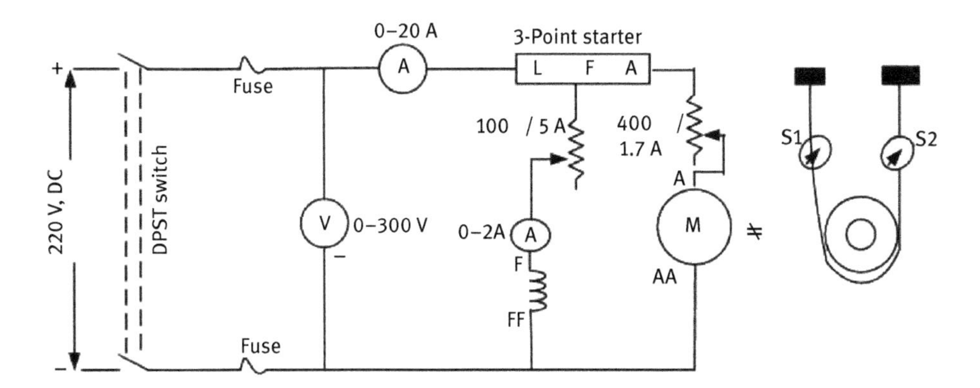

Fig. 2.16: Circuit diagram of DC shunt motor: brake test.

2.9.1.4 Procedure

1. The name plate details of the shunt motor should be noted.
2. The connections of various terminals are connected as per the circuit with Double Pole Single Throw (DPST) off position.
3. Loosen the rope and add water inside the brake drum rim.
4. Switch on the supply (220 V DC) and using 3-point starter get the motor speed to its rated
   value by adjusting armature and field rheostat.
5. Note down the readings of voltmeter, ammeter and balancing weights by changing the load in gradual steps. Maintain terminal voltage and the shunt field current constant.
6. Care must be taken not to exceed the rated value of the motor current.
7. Gradually remove the load and switch off the motor.
8. Calculate various parameters and draw the performance characteristics.

2.9.1.5 Observations

Rated voltage of the motor = Volts rated field current = Amp rated speed = rpm

Table 2.2: Observation table.

S. no	Armature current (I_a) (A)	Speed (N) (rpm)	S_1 (kg)	S_2 (kg)	Power input (P_{in}) (W)	Torque (J/rad)	Speed (ω) (rad/s)	Output power (W)	% η
1									
2									
3									
4									

2.9.1.6 Model graphs

(a) Speed (N) – output power (P), (b) torque (T) – output power, (c) efficiency (η) – output power (P) (d) speed (N) – torque (T).

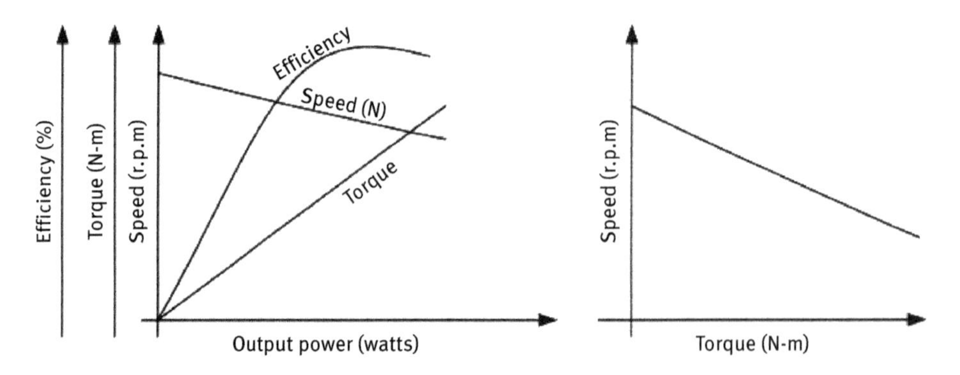

Fig. 2.17: Model graphs of DC shunt motor performance.

2.9.1.7 Practical calculations

Brake test is evaluated on the following rating of the DC shunt motor, and readings were tabulated.

Power = 3.5 kW.
Speed = 1,500 rpm
Voltage = 220 V
Current = 18.7 A
Excitation = 220 V, 1.10 A

Table 2.3: Practical observations.

S. no.	I_L (A)	I_f (A)	$I_a = I_L - I_F$ (A)	N (rpm)	F_1 (kg)	F_2 (kg)	T (Nm)	P_i (W)	P_o (W)	%η
1	2.2	0.58	1.62	1,500	0	0	0	484	0	0
2	3.2	0.58	2.62	1,458	1.2	0.2	1.03	704	157.26	22.33
3	4.2	0.58	3.62	1,423	2.6	0.4	2.26	924	336.77	36.44
4	5.2	0.58	4.62	1,390	4.2	0.8	3.50	1,144	509.46	44.53
5	6.2	0.58	5.62	1,365	5.8	1	4.94	1,362	706.13	51.84

$$\text{Armature current } (I_a) = \text{load current } (I_L) - \text{field current } (I_f)$$

$$\text{Torque, } T = r \times (F_1 - F_2) \times g \text{ Nm}$$

r = radius of the brake drum in meters = 0.105 m.

$$\text{Input power, } P_i = V \times I_L \text{ watts}$$

$$\text{Output power } P_O = \frac{2\pi NT}{60} \text{ watts}$$

$$\%\text{Efficiency} = \frac{P_O}{P_i} \times 100$$

i) Torque $T = r \times (F_1 - F_2) \times g = 0.105 \times (0-0) \times 9.81 = 0$ Nm

Input power, $P_i = V \times I_L = 220 \times 2.2 = 484$ W

$$\text{Output power } P_0 = \frac{2\pi NT}{60} = \frac{2 \times \pi \times 1,500 \times 0}{60} = 0 \text{ Nm}$$

$$\text{Efficiency} = \frac{P_O}{P_i} \times 100 = \frac{0}{100} \times 100 = 0$$

ii) Torque = $0.105 \times (1.2 - 0.2) \times 9.81 = 1.03$ Nm

Input power, $P_i = 220 \times 3.2 = 704$ W

$$\text{Output power, } P_O = \frac{2\pi NT}{60} = \frac{2 \times \pi \times 1458 \times 1.03}{60} = 157.26 \text{ W}$$

$$\%\text{Efficiency, } \eta = \frac{157.26}{704} \times 100 = 22.33\%$$

iii) Torque = $0.105 \times (2.6-0.4) \times 9.81 = 2.26$ Nm

Input power, $P_i = 220 \times 4.2 = 924$ W

$$\text{Output power, } P_O = \frac{2\pi NT}{60} = \frac{2\times\pi\times1423\times2.26}{60} = 336.77 \text{ W}$$

$$\text{Efficiency, } \%\eta = \frac{336.77}{924}\times100 = 36.44\%$$

The following graphs are drawn and shown below:
(a) Speed (N) – output power (P),
(b) torque (T) – output power,
(c) efficiency (η) – output power (P)
(d) speed (N) – torque (T).

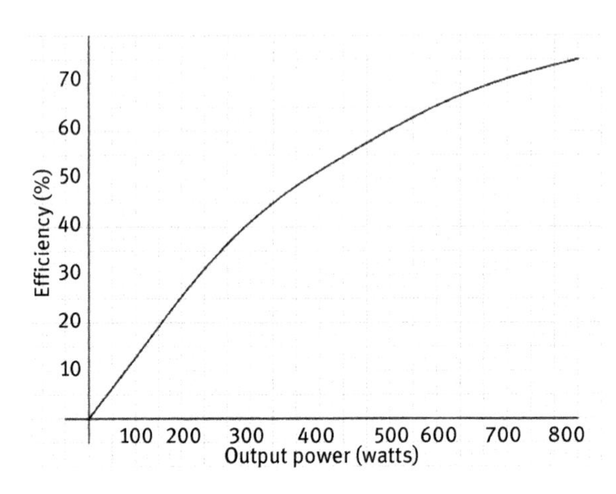

Fig. 2.18: Efficiency versus output power of DC shunt motor.

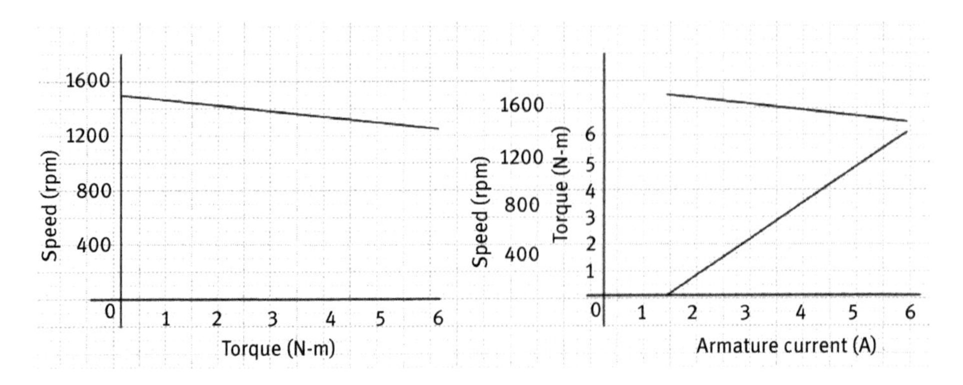

Fig. 2.19: Characteristics of DC shunt motor.

2.9.1.8 Conclusions

The performance characteristics of a DC shunt motor by load test are obtained.

2.9.2 Testing on DC compound motor-brake test

The load test or brake test is conducted on DC compound motor to obtain the performance curves.

2.9.2.1 Apparatus required

Table 2.4: List of required apparatus.

Equipment	Range	Quantity
Ammeter	0–20 A(MC)	1
Ammeter	0–2 A(MC)	1
Voltmeter	0–300 V(MC)	1
Rheostat	400 Ω /1.7 A	1
Tachometer	0–2,000 rpm	1

2.9.2.2 Theory

Brake test is a straight technique of testing the DC machine in which output power, torque, efficiency and speed are determined at different load conditions. A compound motor has a shunt field winding as well as series field winding. If the series field magnetomotive force (MMF) and shunt field MMF help each other, it is a cumulative compound motor. If the series and shunt fields oppose each other, it is a differentially compound motor. The operation of a differential compound motor is unstable. In a cumulatively compounded motor, the fluxes add each other at light loads the shunt field is stronger than series field, so motor behaves shunt motor. At high loads, the series field is stronger than the shunt field, so the characteristics are similar to the series motor.

Power mechanical power developed in the shaft of the motor,

$$P_{out} = T \times \omega \text{ watts} \tag{2.24}$$

where $\omega = 2\pi N/60$

Torque developed at the shaft, T_{shaft} = Force (F) × radius of the drum (r) N-m

$$= (T_1 - T_2)\text{kg} \times 9.8 \times r \tag{2.25}$$

$$\text{Motor output, } P_{out} = T_{shaft} \times \omega = \{(T_1 - T_2) \times 9.8\, r\} \times \{2\pi N/60\} \tag{2.26}$$

where r is the radius of the drum.

Power drawn by the motor (P_{in}) = power drawn by the field + power drawn by the armature

$$P_{in} = V_f I_f + V_a I_a \tag{2.27}$$

$$\% \text{ Efficiency} = [P_{out}/P_{in}] \times 100 \tag{2.28}$$

$$E_b = K\phi\,\omega \tag{2.29}$$

where $\phi = \phi_{sh} + \phi_{se}$

$$\omega = \{1/K\,(\phi_{sh} + \phi_{se})\} \times [V - I_a\,(r_a + r_{se})] \tag{2.30}$$

The induced voltage in the armature,

$$E_b = V - I_a\,R_a - I_{se}R_{se} \tag{2.31}$$

where
V is the applied voltage, I_a is the armature current,
V_f is the voltage to shunt field, I_f is the shunt field current,
I_{se} is the series field current, R_{se} is the series field resistance,
$I_{se} = I_a$ (for a long shunt or a separately excited dc compound motor).

In a separately excited cumulatively compound motor, ϕ_{sh} is constant. Hence, ϕ_{se} increases with an increase in the load. Thus, the speed decreases very quickly in a cumulative compound motor than in a shunt motor:

$$T_a = K\phi\,I_a \tag{2.32}$$

where $\phi = \phi_{sh} + \phi_{se}$ ($\therefore \phi_{sh}$ is constant).

2.9.2.3 Circuit diagrams

Cumulative compound motor

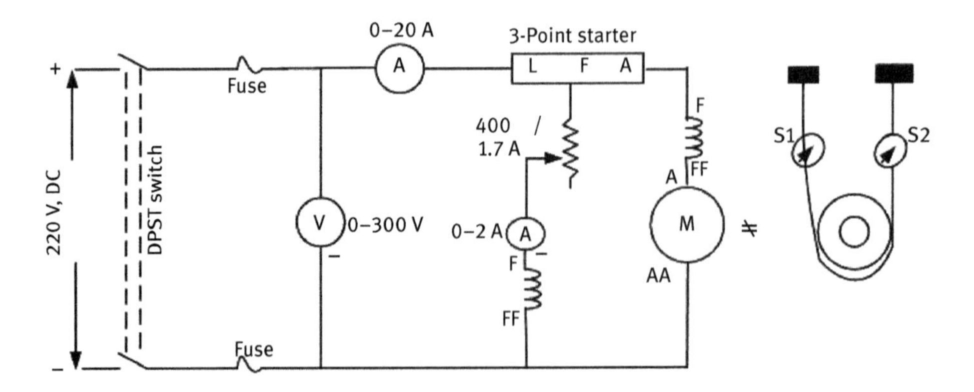

Fig. 2.20: Circuit diagram of DC cumulative compound motor: brake test.

Differential compound motor

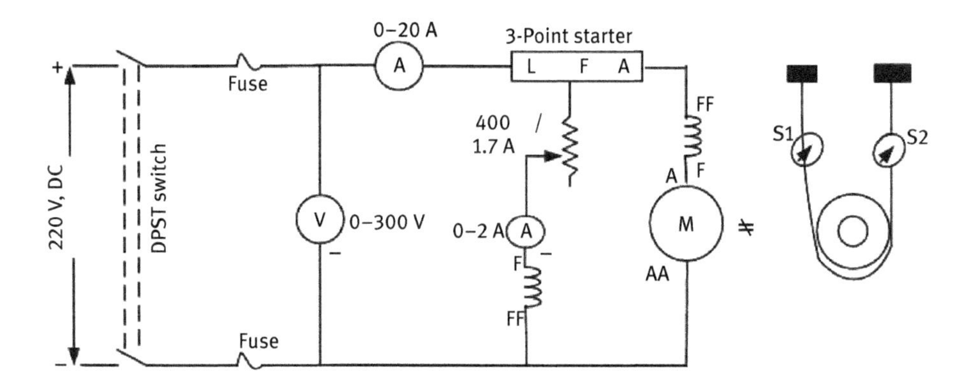

Fig. 2.21: Circuit diagram of DC differential compound motor: brake test.

2.9.2.4 Procedure

1. The name plate details of the shunt motor should be noted.
2. The connections of various terminals are connected as per the circuit with DPST off position.
3. Loosen the rope and add water inside the brake drum rim.
4. Switch on the supply (220 V DC) and using 3-point starter get the motor speed to its rated
   value by adjusting armature and field rheostat.
5. Note down the readings of voltmeter, ammeter and balancing weights by changing the load in gradual steps. Maintain terminal voltage and the shunt field current constant.
6. Care must be taken not to exceed the rated value of the motor current.
7. Gradually remove the load and switch off the motor.
8. Calculate various parameters and draw the performance characteristics.
9. Repeat the experiment for differentially compound motor also.

2.9.2.5 Observation table

Cumulative compound motor

Armature circuit voltage (V_a) = _____ V
Field circuit voltage (V_{sh}) = _____ V
Shunt field current (I_{sh}) = _____ amp

Table 2.5: Observation table for DC cumulative compound motor.

S. no.	Armature current, I_a (A)	Speed, N (rpm)	S1 (kg)	S2 (kg)	Power input, P_{in} (W)	Torque (J/rad)	Speed (ω) (rad/s)	Output power (W)	% η
1									
2									
3									

2.9.2.6 Model graphs

The following graphs are drawn and shown below:

(a) Speed (N) – output power (P)
(b) Torque(T) – Output Power
(c) Efficiency (η) – Output Power (P)
(d) Speed (N) – Torque (T)

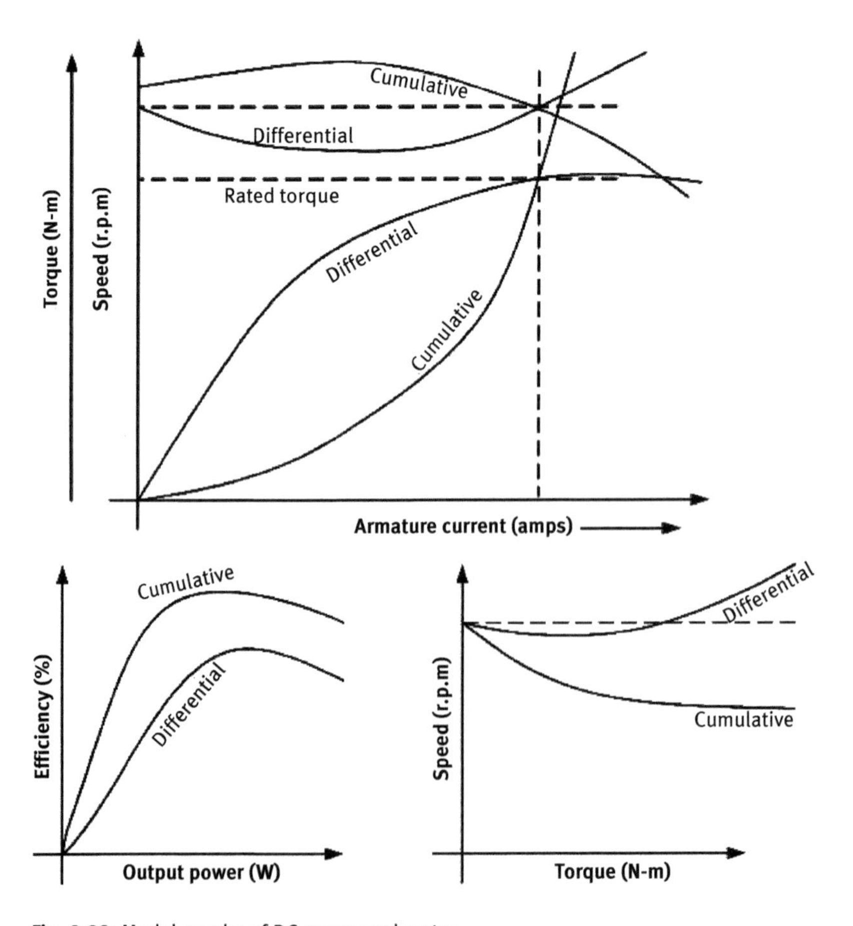

Fig. 2.22: Model graphs of DC compound motor.

2.9.2.7 Practical calculations

The brake test is evaluated on the cumulative compound motor with the following ratings of the motor and readings were tabulated.

Table 2.6: Machine ratings.

Power = 3.7 kW	Current = 19.6 A
Speed = 1,500 rpm	Excitation = 0.56 A
Voltage = 220 V	

Radius of the breakdrum, R = 0.154 m.
Armature resistance R_a = 1.2 Ω.
Series resistance R_{se} = 0.6 Ω.
Shunt resistance R_{sh} = 323 Ω.

Table 2.7: Practical observations.

S. no.	Load current I_L (A)	Field current I_f (A)	Armature current (I_a) (A) $I_a = I_L - I_F$	Speed N (rpm)	Weight F_1 (kg)	Weight F_2 (kg)	Torque (T) (Nm)	Input power (P_I) (W)	Output power P_O (W)	Efficiency %η
1	4	0.6	3.4	1,500	0	0	0	880	0	0
2	5	0.6	4.4	1,450	1	0.2	1.22	1,100	185.15	16.83
3	6	0.6	5.4	1,318	1.2	0.4	1.22	1,320	168.29	12.74
4	7	0.6	6.4	1,210	1.8	0.8	1.53	1,540	193.76	12.58

$$\text{Armature current } (I_a) = \text{load current } (I_L) - \text{field current } (I_f)$$

$$\text{Torque } T = r \times (F_1 - F_2) \times g \text{ N m; } r = \text{radius of the brake drum in meters} = 0.154 \text{ m}$$

$$\text{Input power } P_i = V \times I_L \text{ watts}$$

$$\text{Output power } (P_O) = \frac{2\pi NT}{60} \text{ watts}$$

$$\% \text{ Efficiency} = \frac{P_O}{P_i} \times 100$$

i) $\text{Torque} = 0.154 \times (1.2 - 0.2) \times 9.81 = 0 \text{ Nm}$

$\text{Input power, } P_i = 220 \times 4 = 880 \text{ W}$

$$\text{Power}\,(\text{P}_0) = \frac{2\pi\text{NT}}{60} = \frac{2\times\pi\times1500\times0}{60} = 0\,\text{Nm}$$

$$\%\text{Efficiency},\ \eta = \frac{0}{880}{*}100 = 0\%$$

ii) Torque = 0.154 × (1–0.2) × 9.81=1.20 N m

$$\text{Input power},\ \text{P}_\text{i} = 220\times5 = 1100\,\text{W}$$

$$\text{Output power},\ \text{P}_0 = \frac{2\pi\text{NT}}{60} = \frac{2\times\pi\times1452\times1.2}{60} = 182.21\,\text{W}$$

$$\%\text{Efficiency}, \eta = \frac{182.21}{1,100}\times100 = 16.5\%$$

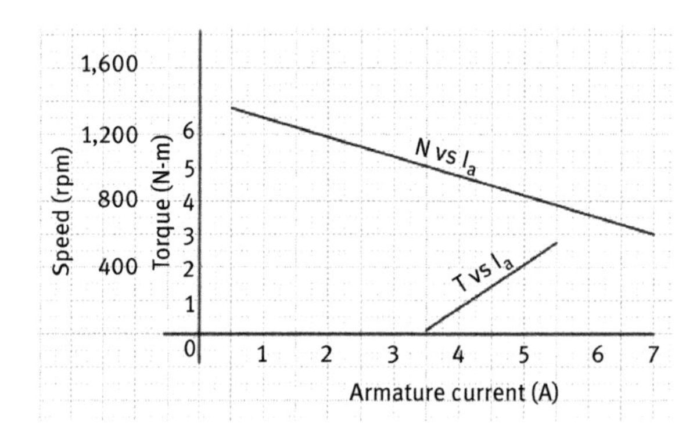

Fig. 2.23: Armature current versus torque and speed curve of DC compound motor.

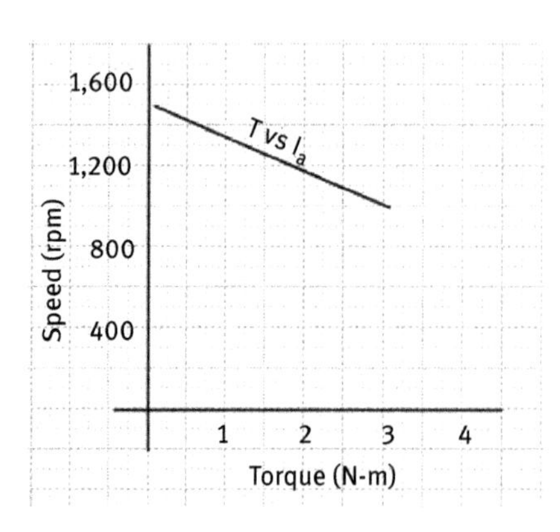

Fig. 2.24: Torque versus speed curve of DC compound motor.

2.9.2.8 Result

The performance curves are attained for DC compound motor by conducting brake test.

2.9.3 Test on DC shunt motor-retardation test

The stray losses are evaluated on DC shunt motor by performing the retardation test.

2.9.3.1 Apparatus required

Table 2.8: List of apparatus required.

Name	Range	Quantity
Voltmeter	0–300 V (MC)	1
Ammeter	0–20 A (MC)	1
Ammeter	0–2 A (MC)	1
Rheostat	400 Ω/1.6 A	1
Rheostat	100 Ω/5 A	1
Tachometer	0–2,000 rpm	1
Stopwatch		1

2.9.3.2 Theory

This method is useful to find stray losses in shunt motors and shunt generators. In this test, the motor rated speed is adjusted above the rated speed and armature supply is disconnected and field winding is excited as it is. Under this condition, we will note down the time taken by the motor to drop the voltage by 25%. After that again the motor speed is increased to the rated value, and field supply is disconnected and armature supply is connected as it is. Under this condition, we will note down the time taken by the motor to drop the voltage by 25%. As a result, the rotor slows down and its kinetic energy is utilized to meet the friction, windage and iron losses (rotational losses).

2.9.3.3 Procedure

1. The name plate details of the shunt motor should be noted.
2. The connections of various terminals are connected as per the circuit with DPST off position.

3. Keep the motor field rheostat (R_{fm}) in the minimum position, armature rheostat in maximum position start the motor by closing the switch and operating the starter slowly.
4. The speed is adjusted to the rated speed by adjusting the field rheostat (or) regulator.
5. The voltage is noted then switch S1 is opened and also note down the time taken to reach the armature voltage to a voltage of 25%, 50%, 75% less than the initial value.
6. Again S1 is closed immediately before the motor reaches to zero speed and rheostats are adjusted until the motor reaches its rated speed.
7. Then S1 is opened and when S2 is closed at this instant, record the readings of ammeter and also note down the time taken to reach the armature voltage to a voltage of 25% less than the initial voltage.

2.9.3.4 Circuit diagram

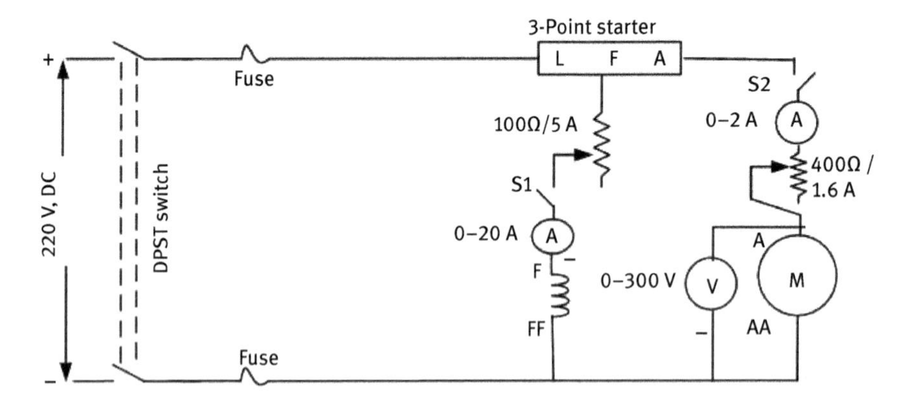

Fig. 2.25: Circuit diagram of DC shunt motor: retardation test.

2.9.3.5 Observation table

S1 is closed and S2 is opened

Table 2.9: Observation table.

S. no.	Voltage when switch S1 is opened, V (V)	Field current, I_f (A)	Time taken to reach the armature voltage to a voltage of 25%, time, T_1 (s)
1			

S1 opened when S2 is closed

Table 2.10: Observation table.

S. no.	Voltage when S1 is opened and at a time S2 is closed (V)	Field current, I_f (A)	Time taken to reach the armature voltage to a voltage of 25%, time, T_2 (s)
1			

2.9.3.6 Practical calculations

Rotational losses or stray losses $P_S = P_{S1}\ (T_2/T_1 - T_2)$; $P_{S1} = V_{avg} \times I_{L,\ avg}$
Input power = $V \times I_L$, I_L = full load current of the motor.
Armature Cu losses = $I_a^2\ R_a$; $I_a = I_L - I_f$
Total losses = armature Cu losses + stray losses
Output power = input power – total losses
Motor efficiency, η = output power/input power

2.9.3.7 Practical calculations

Power = 5.2 kW
Speed = 1,500 rpm
Current = 27.5 A
Voltage = 220 V
Excitation = 0.9 A

S1 is closed and S2 is opened

Table 2.11: Practical observations.

S. no.	Voltage when switch S_1 is opened, V (V)	Field current, I_f (A)	Time taken to reach the armature voltage to a voltage of 25%, time, T_1 (s)
1	165	5	7

S1 is opened when S2 is closed

Table 2.12: Practical observations.

S. no.	Voltage when S_1 is opened and at a time S_2 is closed (V)	Field current (I_f) (A)	Time taken to reach the armature voltage to a voltage of 25%, time, T_2 (s)
1	165	5	4

$$\text{Input power } (P_{S1}) = V \times I_L = 165 \times 5 = 825 \text{ W}$$

If the change in both cases is same, then $P_S = P_{S1} \times \frac{T_2}{(T_1 - T_2)} = 825 \times \frac{4}{(7-4)} = 1,100$ W

Conclusions

Retardation test is performed and stray losses of the given DC shunt machine are found.

2.9.4 Swinburne's test

The efficiency is calculated by conducting Swinburne's or no-load test on DC shunt motor.

2.9.4.1 Apparatus required

Table 2.13: List of required apparatus.

Name	Range	Quantity
Voltmeter	0–250 V (MC)	1
Voltmeter	0–30 V (MC)	1
Ammeter	0–5 A (MC)	1
Ammeter	0–2 A (MC)	1
Tachometer	0–2,000 rpm	1
Rheostat	100 Ω/5 A	1
Rheostat	400 Ω/1.7 A	1

2.9.4.2 Theory

The no-load or Swinburne's test is conducted on DC shunt motor. It is an indirect method of testing. In this test, no load losses of the shunt motor are measured. As the test is conducted on load, the power requirement is less. Therefore, this method is very useful to predetermine the efficiency of large size generators and motors. But the major drawbacks are efficiency calculated without actual load, and temperature effect is not considered.

Efficiency of a DC machine

No load rotational loss, $W_o = V_t \times I_a - I_a^2 \times R_a$ watts

Shunt field loss = $V_t \times I_f$ watts

$I_a = I_L - I_f$ for motor

$I_a = I_L + I_f$ for generator

Motor input = $V_t \times I_L$ watts
Copper loss = $I_a^2 R_a$ watts

$$\text{Efficiency, } \%\eta_m = 1 - \frac{W_0 + I_a^2 R_a + I_f V_t}{V_t \times I_L} \times 100$$

2.9.4.3 Circuit diagram

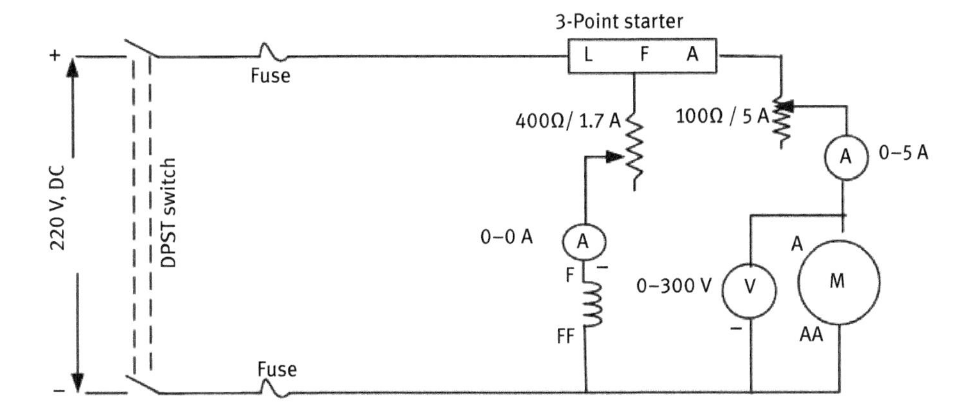

Fig. 2.26: Circuit diagram of DC shunt motor: Swinburne's test.

Armature resistance (R_a)

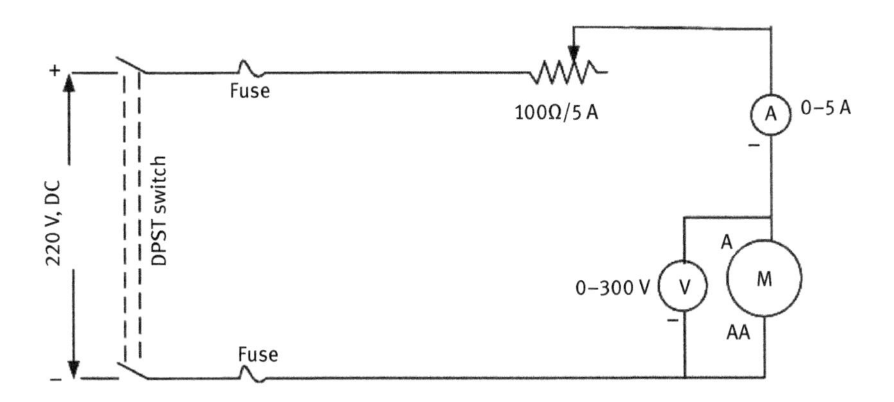

Fig. 2.27: Circuit diagram of DC shunt motor: armature resistance evaluation.

2.9.4.4 Procedure

1. The name plate details of the shunt motor should be noted.
2. The connections of various terminals are connected as per the circuit with DPST off position.
3. Set the field rheostat of the motor at minimum resistance position and motor field rheostat in maximum resistance position.
4. Switch on the supply (220 V DC) and using 3-point starter get the motor speed to its rated value by adjusting armature and field rheostat.
5. Note down the readings of voltmeter, ammeter readings.
6. Measure the field resistance and armature resistance.
7. Do the required calculations to find the efficiency as motor and generator.
8. Plot the graphs between output power versus efficiency of the machine as a generator and as a motor.

2.9.4.5 Observations

Table 2.14: Observation table.

S. No.	Supply voltage (V)	Speed, N (rpm)	Armature current, I_a (A)	R = V/I (ohms)
1				

Armature resistance (Ra)

Table 2.15: Observation table.

S. No.	Supply voltage (V)	Speed, N (rpm)	Armature current, I_a (A)	Field current, I_f (A)
1				

At no-load

Speed (N) =______rpm

Armature voltage =______V

Field voltage =______V

Armature current =______A

Field current =______A

Armature resistance =______Ω

Field resistance =______Ω

As a motor

Table 2.16: Observation table.

S. No.	Load current, I_L (A)	Field current, I_f (A)	Armature current (Ia) Ia = I_L–I_f (A)	Supply voltage, V (V)	Speed, N (rpm)
1					

2.9.4.6 Model graph

The test is conducted separately for generator and motor at various load conditions and calculated efficiency. The graph is drawn output power verses efficiency.

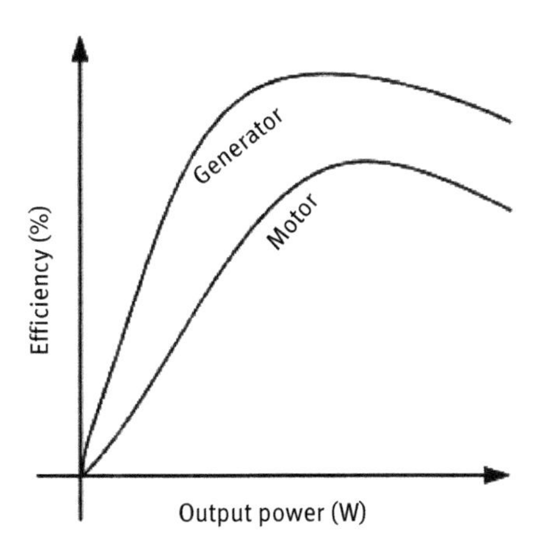

Fig. 2.28: Model graph DC shunt machine: efficiency versus output power.

2.9.4.7 Practical calculations

The Swinburne's test is conducted on DC shunt machine with the following ratings and readings tabulated.

Power = 3.5 kW Current = 21 A
Speed = 1,500 rpm Self-excitation = 0.5 A
Voltage = 220 V Armature resistance $R_a = 2\ \Omega$

Table 2.17: Practical observations.

S. no.	Supply voltage, V (V)	Speed, N (rpm)	Load current, I_L (A)	Field current, I_f (A)
1	220	1,500	2.2	0.5

Armature current (I_{a0}) = load current (I_L) – field current (I_f) = 2.2 – 0.5 = 1.7 A
terminal voltage V_t = 220 V, armature resistance, R_a = 2.

At full load

No load rotational loss, $W_o = V_t \times I_a - I_a^2 \times R_a = 220 \times 1.7 - (1.7)^2 \times 2 = 368.22$ watts

Shunt field loss $= V_t \times I_f = 220 \times 0.5 = 110$ watts

$I_a = I_L - I_f = 21 - 0.5 = 20.5\ \text{A}$

Motor input $= V_t \times I_L = 220 \times 21 = 4,620\ \text{W}$

Copper loss $= I_a^2 R_a = (20.5)^2 \times 2 = 840.5\ \text{W}$

$$\text{Efficiency, } \%\eta_m = 1 - \frac{W_0 + I_a^2 R_a + I_f V_t}{V_t \times I_L} \times 100 = 1 - \frac{368.22 + 840.5 + 110}{4620} \times 100 = 71.46\%$$

At $\frac{3}{4}$ load

$$\text{Load current } (I_L) = 21 \times \frac{3}{4} = 15.75\ \text{A}$$

$$\text{Armature current } (I_a) = I_L - I_F = 15.75 - 0.5 = 15.25\ \text{A}$$

$$\text{Copper losses} = I_a^2 R_a = (15.25)^2 \times 2 = 465.125\ \text{W}$$

No load rotational loss $W_0 = 368.22$, Shunt field loss $= V_t I_f = 110$ W

$$\text{Motor input} = V_t \times I_L = \frac{220 \times 21 \times 3}{4} = 3,465\ \text{W}$$

$$\text{Motor efficiency } \%\eta_m = 1 - \frac{368.22 + 465.125 + 110}{3,465} \times 100 = 72.78\%$$

At $\frac{1}{2}$ load

$$\text{Load current } (I_L) = 21 \times \frac{1}{2} = 10.5\ \text{A}$$

$$\text{Armature current } (I_a) = I_L - I_F = 10.5 - 0.5 = 10\ \text{A}$$

$$\text{Copper losses} = I_a^2 R_a = 10^2 \times 2 = 200$$

$$\text{Motor input} = V_t \times I_L = \frac{220 \times 21}{2} = 2,310\ \text{W}$$

$$\text{Efficiency } \%\eta_m = 1 - \frac{368.22 + 200 + 110}{2,310} \times 100 = 70.64\%$$

At $\frac{1}{4}$ load

Load current $(I_L) = 21 \times \frac{1}{4} = 5.25\ A$

Armature current $(I_a) = I_L - I_F = 5.25 - 0.5 = 4.75\ A$

Copper loss $= I_a^2 R_a = (4.75)^2 \times 2 = 45.125\ W$

Motor input $= V_t \times I_L = 220 \times 5.25 = 1,155\ W$

$$\%\eta_m = 1 - \frac{368.22 + 45.125 + 110}{1,155} \times 100 = 54.69\%$$

At no load

$I_{a0} = 1.7\ A$

$I_{a0}^2 R_a = (1.7)^2 \times 2 = 5.78\ W$

Motor input, $V_t I_{L0} = 220 \times 2.2 = 484\ W$

$$\%\eta_m = 1 - \frac{368.22 + 5.78 + 100}{484} \times 100 = 3\%$$

The above calculations can be done for generator by considering $I_a = I_L + I_{sh}$

Table 2.18: Efficiency calculated as per load.

S. no.	Load	Efficiency (%)
1	0	3
2	1/4	54.69
3	1/2	70.64
4	3/4	72.78
5	1	71.46

Result: The efficiency of a DC shunt motor and generator is calculated by conducting Swinburne's test and plotted curve between efficiency versus output power.

2.9.5 Testing on DC shunt motor-generator set – Hopkinson's test

This test is conducted on two identical DC shunt machines to evaluate their efficiencies at various loads. It is also known as back-to-back test.

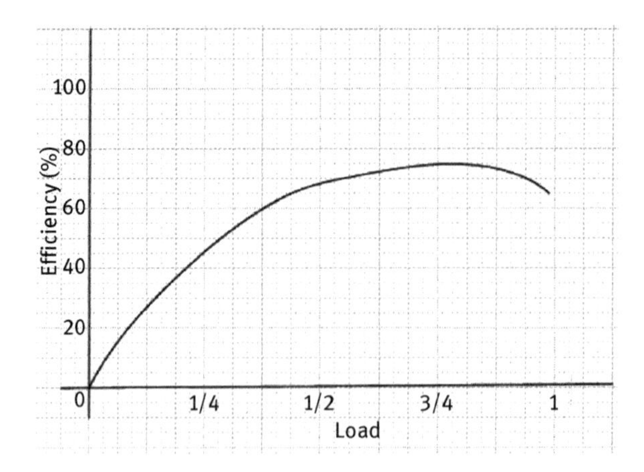

Fig. 2.29: Efficiency versus load curve of DC shunt motor.

2.9.5.1 Apparatus required

Table 2.19: List of required apparatus.

Name	Range	Quantity
Voltmeter	0–300 V (MC)	1
Voltmeter	0–500 V (MC)	1
Voltmeter	0–30 V (MC)	1
Ammeter	0–5 A (MC)	1
Ammeter	0–20 A (MC)	1
Rheostat	400 Ohms/1.7A	2
Rheostat	100 Ohms/5A	1
Ammeter	0–2 A (MC)	2
Tachometer	0–2,000 rpm	1

2.9.5.2 Theory

Hopkinson's back-to-back test is conducted on identical motor-generator set. Initially, the motor is started by the mains. Once the motor reaches to the rated speed, the generator power is connected to the motor and start running without taking the power from the mains but it will take the power to meet the motor losses only. Therefore, the Hopkinson test is also termed as regenerative test. Due to the losses, the generator output power is not enough to run the motor and vice versa. Thus, these losses in the machines are provided electrically from the mains.

2.9.5.3 Circuit diagram

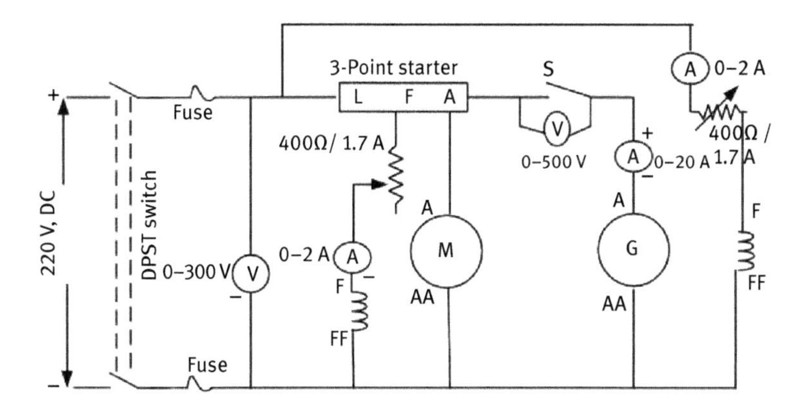

Fig. 2.30: Circuit diagram of DC shunt machines: Hopkinson's test.

Armature resistance (R_a)

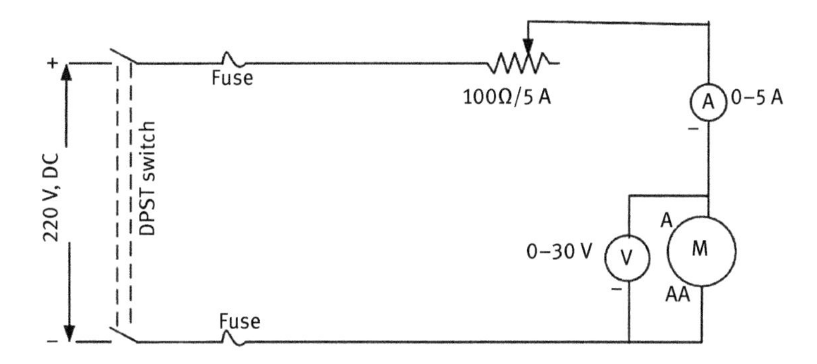

Fig. 2.31: Circuit diagram of DC shunt generator: armature resistance evaluation.

2.9.5.4 Procedure

1. The name plate details of the shunt motor should be noted.
2. The connections of various terminals are connected as per the circuit with DPST off position.
3. Set the field rheostat of the motor at minimum resistance position and motor field rheostat in maximum resistance position and switch "S" in open position.
4. Switch on the supply (220 V DC) and using 3-point starter get the motor speed to its rated value by adjusting armature and field rheostat.
5. Now slowly excite the generator field and confirm the voltmeter across the switch is zero, then close the switch "S."

6. Record the meter readings.
7. Open the switch "S" and slowly reduce the excitation.
8. Switch off the mains.

Procedure to calculate the efficiencies

I_1 = Motor input current drawn from the mains
I_2 = Current drawn by the motor from the generator
I_{f1} = Motor field current.
I_{f2} = Generator field current.
R_{a1} and R_{a2} = Motor and generator armature resistance
Total current supplied to the motor = $I_1 + I_2$

If "V" is the supply voltage connected across the motor and "V_f" is the voltage applied across the field.
Motor input power, $P_{in} = V \times (I_1 + I_2)$
Generator output power, $P_O = V \times I_1$
Generator copper losses = $I_2^2 \times R_{a2}$
Motor copper losses = $(I_1 + I_2)^2 \times R_{a1}$
Generator shunt copper losses = $V_f \times I_{f2}$
Motor shunt copper losses = $V_f \times I_{f1}$
The power supplied by mains is equal to losses in the generator and motor.
Power drawn from supply = V_{I1}
Stray losses of both the machines = $V_{I1} - [I_2^2 R_{a2} + (I_1 + I_2)\ R_{a1}] = W_{TS}$
Therefore, the total stray losses for each set $W_S = W_{TS}/2$
In this test, the stray losses are considered equal in generator and motor.
The stray losses for each machine = W_S

For generator

Total losses, $W_g = I_2^2 R_{a2} + I_{f2} V_f + Ws$
Output power = $V \times I_2$
Therefore, % efficiency $\eta_g = (V \times I_2/(V \times I_2 + W_g)) \times 100$

For motor

Total losses, $W_m = (I_1 + I_2)^2 \times R_{a1} + V_f \times I_{f1}$
Input power, $P_{in} = V(I_1 + I_2) + V_f \times I_{f1}$
Output power, $P_o = P_{in} - W_m$
Therefore, % efficiency, $\eta_m = (P_O/P_{in}) \times 100$

2.9.5.5 Observations table

The generator and motor armature resistance $R_{a1} = R_{a2}$ = ______ ohms
Applied voltage, V =______volts, voltage applied to the field V_f =______volts

Table 2.20: Observation table.

S. no.	Input voltage (V)	Input current, I_1 (A)	Generator armature current $I_a = I_2$ (A)	Generator field current $I_f = I_3$ (A)	Motor field current $I_f = I_4$ (A)
1					

Efficiency calculation

For motor

Table 2.21: Observation table.

S. no.	Motor input power (W)	Motor armature Cu loss (W)	Motor field Cu loss (W)	Stray loss (W)	Total losses of motor (W)	Output of motor (W)	η of motor
1							

For generator

Table 2.22: Observation table.

S. no.	Generator output (W)	Generator armature Cu loss (W)	Generator field Cu loss (W)	Stray loss (W)	Total losses of generator (W)	Input of generator (W)	η of generator
1							

2.9.5.6 Practical calculations

The Hopkinson's test is conducted on the following rating of the machine, and readings were tabulated;

Table 2.23: Machine ratings.

	Generator	Motor
Rated power	5.2 kW	5.2 kW
Rated speed (N)	1,500 rpm	1,500 rpm
Applied voltage (V)	220 V	220 V
Full load current	23.6 A	27.5 A
Excitation voltage	1.65 A	0.9 A

Table 2.24: Practical observations.

S. no.	Input voltage (V)	Input current, I_1 (A)	Generator armature current $I_a = I_2$ (A)	Generator field current $I_f = I_3$ (A)	Motor field current $I_f = I_4$ (A)
1	220	2	2.5	0.46	0.8

Armature resistance $R_a = 1.3\ \Omega$.

For motor

$$\begin{aligned} \text{Armature current} I_a = I_1 + I_2 & = 2 + 2.5 = 4.5\text{ A} \\ \text{Armature Culosses } = R_a \times (I_1 + I_2)^2 & = 1.3 \times (2 + 2.5)^2 = 26.325\text{ W} \\ \text{Motor input power } = V \times (I_1 + I_2) & = 220 \times (2 + 2.5) = 990\text{ W} \end{aligned}$$

For generator

$$\text{Armature current} = I_2 + I_4 \ = 2.5 = 0.8 = 3.3\ \text{A}$$

$$\text{Armature copper losses} = (I_1 + I_2)^2 \times R_a = (2.5 + 0.8)^2 \times 1.3 = 4.29\ \text{W}$$

Stray losses (w) = friction and windage losses for two machines

$$= V\ (I_1 + I_3)^2\ R_a - (I_2 + I_4)^2\ R_a - R_a\ (I_1 + I_2)^2 - VI_3 - VI_4\ = 333.518\ \text{W}$$

Stray losses for each machine = W/2 = 166.759 watts

Motor losses (W_m) = armature copper losses + shunt copper losses + w/2

$$= R_a\ (I_1 + I_2)^2 + VI_3 + w/2 = (1.3) \times (4.5)^2 + 220 \times 0.46 + 166.759 = 333.514\ \text{W}$$

Motor efficiency

$$\%\eta_m = \frac{\text{Input power} - \text{Wm}}{\text{Input power}} \times 100 = \frac{990 - 333.514}{990} \times 100 = 66.3\%$$

Generator efficiency

$$\text{Generator output} = V \times I_2 = 220 \times 2.5 = 550\ \text{W}$$

Generator losses, W_g = armature copper losses + shunt copper losses + W/2

$$= R_a\ (I_4 + I_2)^2 + VI_4 + W/2 = 356.916\ \text{W}$$

$$\text{Generator efficiency, } \%\eta_g = \frac{\text{output power}}{\text{input power } + \text{ Wg}} \times 100 = \frac{550}{550 + 356.916} \times 100 = 60.64\%$$

Result: Performed the Hopkinson's test on two similar DC shunt machines, and obtained their efficiencies at various loads and drawn graphs.

2.9.6 Field test

This test is conducted on two identical DC series machines to evaluate their efficiencies at various loads.

2.9.6.1 Apparatus required

Table 2.25: List of required apparatus.

Name	Range	Quantity
Voltmeter	0 – 250 V (MC)	2
Ammeter	0 – 20 A (MC)	2
Rheostat	100 Ω/5 A	1
Resistive load	0–3 kW	1
Tachometer	0–2,000 rpm	1

2.9.6.2 Theory

This test is applicable for two series machines that are coupled mechanically. Series machines cannot be tested on no-load conditions due to dangerously high speeds. One machine normally runs as motor and drives generator whose output is wasted in a variable load R. The fields of two machines are connected in series to make iron losses of both the machines equal. Field test is conducted on series machines to obtain its efficiency. In this test, one machine acts as motor and another as generator.

Calculations

I_1 = The current drawn by the motor from the mains
V_1 = Voltage applied across the motor terminals
I_2 = Load current
V_2 = Load voltage
Power drawn from the supply (P_{in}) = $V_1 \times I_1$ watts
Power consumed in the load (P_{out}) = $V_2 \times I_2$ watts
Total losses in the two machines (W_L) = $(V_1 \times I_1 - V_2 \times I_2)$ watts
Total copper losses in the two machines (W_{Cu}) = $I_1^2 (R_{a1} + R_{Se1} + R_{Se2}) + I_2^2 R_{a2}$
Stray losses in two machines = Total losses (W_L) – Cu losses(W_{Cu})
The stray losses for each machine = $W_{Stray} = (W_L - W_{Cu})/2$
Hence, the stray losses are same in both the machines.

For motor

Motor power input, $P_{in} = V_1 \times I_1 - I_1^2 R_{Se2}$

Total losses in the motor, $W_{ml} = W_S + I_1^2(R_{a1} + R_{Se2})$

Motor output, $P_{out} = P_{in} - W_{ml}$

% Efficiency of the motor η_m = Motor output (P_{out})/motor input (P_{in}) ×100

For generator

Power input of the generator $P_{out} = V_2 \times I_2$

Total losses in the generator $W_{gl} = W_S + I_1^2 R_{Se2} + I_2^2 R_{a2}$

Input to the generator, $P_{in} = P_{out} + W_{gl}$

$$\%\text{Efficiency of the generator, } \eta_g = \frac{\text{generator output}}{\text{generator input}} \times 100$$

2.9.6.3 Circuit diagram

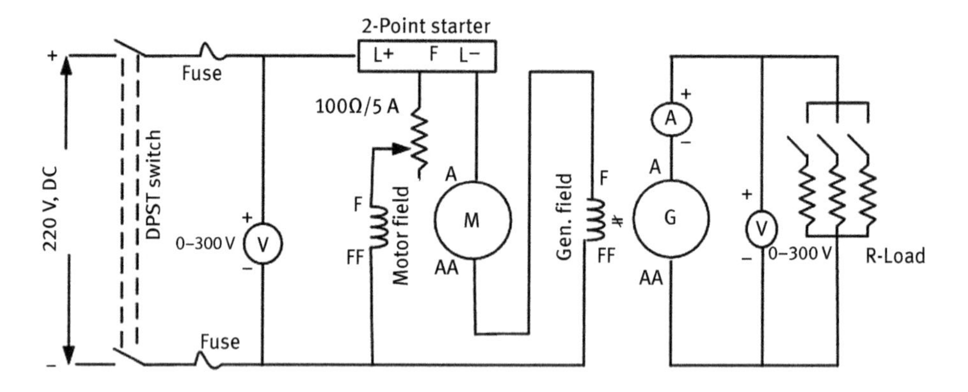

Fig. 2.32: Circuit diagram of DC series machines: field test.

Armature resistance (R_a)

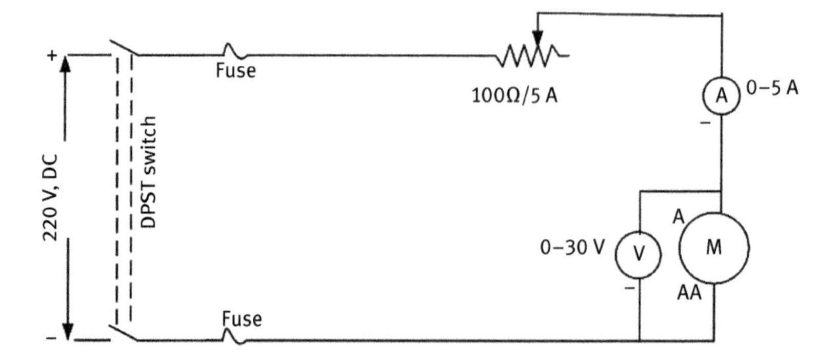

Fig. 2.33: Circuit diagram of DC series machine: armature resistance evaluation.

2.9.6.4 Procedure

1. The name plate details of the shunt motor should be noted.
2. The connections of various terminals are connected as per the circuit with DPST off position.
3. Set the field rheostat of the motor at minimum resistance position and motor field rheostat in maximum resistance position.
4. Switch on the supply (220 V DC) and using 2-point starter get the motor speed to its rated value by adjusting armature rheostat.
5. Now slowly increase the load up to the rated value gradually on the generator.
6. Record the meter readings at different loads.
7. Decrease the terminal voltage slowly and switch off the mains.

2.9.6.5 Observations table

Armature resistance of the motor (R_{a1}) =________Ω
Series field resistance of the motor (R_{Se1}) =________Ω
Armature resistance of the generator (R_{a2}) =________Ω
Series field resistance of the generator (R_{Se2}) =________Ω

Table 2.26: Observation table.

S. no.	Load (kW)	Supply voltage, V_1 (V)	Current drawn from the supply, I_1 (A)	Load voltage, V_L (V)	Load current, I_2 (A)	Motor efficiency, % η_m	Generator efficiency, $\%\eta_m$
1							
2							
3							
4							

2.9.6.6 Practical calculations

The field test is conducted on two series machines with the following rating and the readings were tabulated.

Table 2.27: Machine ratings.

Generator	Motor
Power = 3.7 kW	Power = 5.2 kW
Speed = 1,500 rpm	Speed = 1,500 rpm
Voltage = 220 V	Current = 27.5 A
Current = 19.5 A	Voltage = 220 V

Table 2.28: Practical observations.

S. no.	Load (kW)	Supply voltage, V_1 (V)	Current drawn from the supply, I_1 (A)	Load voltage, V_L (V)	Load current, I_2 (A)	Motor efficiency, % η_m	Generator efficiency, %η_m
1	1	40.2	10.9	115	0.4	37.64	27.88
2	2	47.5	11.5	130	0.8	43.85	30.81
3	3	54.5	12.8	145	1.4	49.46	58.82
4	4	60	13.6	155	1.9	53.59	67.33

For load 1 A

For motor

Motor input $= V \times I_1 = (40.2) \times (10.9) = 438.18$ W

$$W_t = V_1 \times I_1 - V_2 \times I_2 = (40.2) \times (10.9) - (115) \times (0.4) = 392.18 \text{ W}$$

$$W_c = I_1^2(R_a + 2R_{se}) + I_2^2 R_a = (10.9)^2 \times (1.3 + (2 \times 0.6)) + (0.4)^2 \times (1.3) = 297.23 \text{ W}$$

$$W_{stray} = W_t - W_c = 392.18 - 297.233 = 94.97 \text{ W}$$

Motor losses, $W_m = I_1^2(R_a + R_{se}) + W_S/2 = (10.9)^2 \times (1.3 + 0.6) + 94.947/2 = 273.21$ W

$$\text{Motor efficiency } \eta_m = \frac{V_1 I_1 - W_m}{V_1 I_1} = \frac{438.180 - 273.2125}{438.18} \times 100 = 37.64\%$$

For generator

Generator output $= V_2 \times I_2 = 115 \times 0.4 = 46$ W

Field copper losses $= I_2^2 R_a = 0.4^2 \times 1.3 = 0.208$ W

Total losses, $W_g = I_2^2 R_a + I_1^2 R_a + W_S/2 = (0.4)^2 \times (1.3) + 10.9^2 \times 0.6 + 94.947/2 = 118.9675$ W

$$\text{Generator efficiency } \eta_G = \frac{V_2 I_2}{V_2 I_2 + W_g} = \frac{46}{46 + 118.9675} \times 100 = 27.88\%$$

2.9.6.7 Result

Perform the field test and determine the efficiency of the two given DC series machines that are mechanically coupled.

2.10 Problems

1. The load test is conducted on a DC shunt motor and the following results were obtained:
   Balancing weight readings: 10 and 35 kg   Diameter of the drum: 40 cm
   Speed: 950 rpm   Supply voltage: 200 V
   Line current: 30 A
   Calculate the output power and the efficiency.

Solution

Force on the drum surface $F = (W_1 - W_2)\ g = (35 - 10) = 25$ kg, wt $= 25 \times 9.8$ N
Drum radius R= 20 cm = 0.2 m
Torque, $T_{sh} = F \times R = 25 \times 9.8 \times 0.2 = 49$ N
$N = 950/60 = 95/6$ rps, $\omega = 2\pi\ (95/6) = 99.5$ rad/s
Motor output = $T_{sh} \times \omega$ watt $= 49 \times 99.5 = 4{,}876$ W
Motor input = applied voltage × full load current ($V \times I$) = $200 \times 30 = 6{,}000$ W

$$\text{Efficiency} = \frac{\text{motor output}}{\text{motor input}} = \frac{4,876}{6,000} = 0.813 \text{ or } 81.3\%$$

---

2. The retardation test is conducted on a DC motor, the induced emf in the armature falls from 220 to 190 V in 30 s on disconnecting the armature from the supply. The same drop takes place in 20 s if instantly after disconnection, and armature is connected to a resistance that takes 10 A during this fall. Find stray losses of the motor.

Solution

Let W = stray losses
Average voltage across resistance = (200 + 190)/2 = 195 V, average current = 10 A
Therefore, the power absorbed (W') = V × I = 195 × 10 = 1,950 W

$$\text{Using the relation } W = W'\left(\frac{T_2}{T_1 - T_1}\right) = 1,950 \times \left(\frac{20}{30 - 20}\right) = 3,900 \text{ W}$$

---

3. In a test on a DC shunt generator whose full-load output is 200 kW at 250 V, the following data were obtained:

When running light as a motor at full speed, the line current was 36 A, the field current was 12 A and the supply voltage was 250 V. Obtain the efficiency of the generator at full load and half load. Neglect brush voltage drop.

Solution

At no-load

$I_a = I_L - I_f = 36 - 12 = 24$ A; $R_a = 6/400 = 0.015\ \Omega$

$\therefore$ Armature Cu loss $= I_a^2 \times R_a = 24^2 \times 0.015 = 8.64$ watt

No-load input = total losses in machine $= V \times I_a = 250 \times 36 = 9{,}000$ W

Constant losses = No-load input – armature Cu loss =9,000–8.64 = 8,991.4 W

At full load

Output power = 200,000 W; output current = 200,000/250 = 800 A; $I_{sh} = 12$ A

$\therefore$ F.L. armature current $I_L + I_{sh} = 800 + 12 = 812$ A

$\therefore$ F.L. armature Cu losses $= 812^2 \times 0.015 = 9{,}890$ W

$\therefore$ F.L. total losses = 9,890 + 8,991.4 = 18,881 W

$$\text{Efficiency, } \%\eta = \frac{\text{output power}}{\text{input power}} \times 100 = \frac{200{,}000}{200{,}000 + 18{,}881} \times 100 = 91.4\%$$

At half load

Output power = 100,000 W; output current = 100,000/250 = 400 A

$$I_a = 400 + 12 = 412\ \text{A}$$

$$I_a^2 R_a = 412^2 \times 0.015 = 2{,}546\ \text{W}$$

Total losses = 8,991.4 + 2,546 = 11,537 W

$$\therefore \text{Efficiency, } \%\eta = \frac{\frac{1}{2}(200{,}000)}{\frac{1}{2}(200{,}000) + 11{,}537} \times 100 = 89.6\%$$

---

4. Hopkinson's test is conducted on two shunt machines. It draws 15 A at 200 V from the supply. The motor current is 100 A and the shunt field currents are 3 and 2.5 A. If the armature resistance of each machine is 0.05 Ω, calculate the efficiency of each machine.

Solution

Current drawn by the armature of the machine = 15 A,

The motor current = 100 A

Armature copper loss of the motor $= I_{am}^2 R_a = 100^2 * 0.05 = 500$ W

The generator armature–current $I_m - I_L = 100 - 15 = 85$ A

Armature copper losses of the generator $= I_{ag}^2 R_a = 85^2 * 0.05 = 361$ W

For each machine, no load losses (mechanical, core and stray losses)

$$= 1/2\ (VI_a - I^2_{am}\ r_{am} - I^2_{ag}\ r_{ag})$$

$$= 1/2\,(200 \times 15 - 100^2 \times 0.05 - 85^2 \times 0.05)$$

$$= 1/2\,(3,000 - 500 - 361) = 1,069.5\text{W} \cong 1.07\text{ kW}$$

Shunt field copper–losses of the motor = 200 × 3 = 600 W
Shunt field copper–losses of the generator = 200 × 2.5 = 500 W
The total losses in the motor = 600 + 1,069.5 + 500 = 2,169.5 W
The total losses in the generator = 500 + 1,069.5 + 361 = 1,931 W

$$\text{Efficiency of motor} = \frac{\text{motor output}}{\text{motor input}} \times 100$$

Motor input:

(i) 200 × 100 = 20 kW to armature of the motor
(ii) 2000.6 kW to field winding of the motor

Total power input to the motor = 20.6 kW
From armature side, losses to be catered are:

(i) Stray losses + no load mech. losses + core losses = 1.07 kW
(ii) Armature copper loss = 0.5 kW motor output from armature
= 20 − 0.5 − 1.07 = 18.43 kW

$$\text{Motor efficiency} = \frac{18.43}{20.6} \times 100 = 89.47\%$$

Generator armature output = 200 × 85 = 17 kW
Generator losses:

(a) Field winding: 0.5 kW
(b) Total no-load losses: 1.07 kW
(c) Armature copper losses: 0.36 kW

Total losses in generator = 1.93 kW

$$\text{Generator efficiency} = \frac{17}{17 + 1.93} \times 100 = 89.80\%$$

---

5. A field test is conducted on identical series motor, gave the following results:

Motor

Current through the armature = 56 A
Voltage across the armature = 590 V
series field voltage drop = 40 V

Generator

Current through the armature = 44 A
Voltage across the armature = 400 V
Series field voltage drop = 40 V
Resistance of motor and generator armature = 0.3 Ω
Calculate the efficiency of the motor and generator.

Input voltage to motor = armature voltage + voltage drop across motor

= 590 + 40 = 630 V

Total input power to motor = Motor input voltage × armature current = 630 × 56

= 35,280 W

Generator output power = armature voltage × current = 400 × 44 = 17,600 W

Total losses in the two machines are = total power input to motor − generator output power

= 35,280 − 17,600 = 17,680 W

Series field resistance, $R_{se} = 40/56 = 0.714\ \Omega$

Armature and field Cu losses, $W_{cu} = (R_a + 2R_{se})\, I_1^2 + I_2^2 R_a$

$(0.3+2\times0.714)56^2 + 44^2\times0.3 = 5{,}425 + 581 = 6{,}006$ W

$$\text{Stray losses for the set} = W_t - W_{cu} = 17,680 - 6,006 = 11,674\ \text{W}$$

$$\text{Stray losses per machine, } W_s = \frac{W_t - W_{cu}}{2} = \frac{11,674}{2} = 5,837\ \text{W}$$

Motor efficiency

Motor armature input, P_{in} = armature voltage × motor current = 590 × 56 = 33,040 W

Total losses in the motor = armature + field Cu losses + stray losses

$$= W_{ml} = W_S + I_1^2(R_a + R_{Se}) = 5,837 + 56^2(0.3 + 0.714) = 5,837 + 3,180 = 9,017\ \text{W}$$

Motor output, $P_{out} = P_{in} - W_{ml} = 33,040 - 9,017 = 24,023$ W

$$\%\text{Efficiency of the motor, } \eta_m = \frac{\text{motor output}}{\text{motor input}} \times 100 = \frac{24,023}{33,040} \times 100 = 72.2\%$$

Generator efficiency

Power output of the generator (P_{out}) = $V_2 \times I_2$ = 400 × 44 = 17,600 W

Total losses in the generator, $W_{gl} = W_S + (I_1^2 R_{Se2} + I_2^2 R_{a2})$

$= 5{,}837 + 56^2 \times 0.714 + 44^2 \times 0.3$

$= 5{,}837 + 2{,}239.1 + 580.8 = 8{,}658$ W

Input to the generator $P_{in} = P_{out} + W_{gl} = 17{,}600 + 8{,}658 = 26{,}258$ W

$$\%\text{Efficiency of the generator, } \eta_g = \frac{\text{generator output}}{\text{generator input}} \times 100 = \frac{17,600}{26,258} \times 100 = 67\%$$

2.11 Viva-Voce questions

Brake test on DC shunt motor

1. What do you understand by the term “Back EMF?”
2. Mention the applications of DC shunt motor.
3. The shunt motor is a constant speed motor. Justify.

4. What is the condition for maximum efficiency of the motor?
5. What are the effects of commutation?
6. Define reactance voltage.
7. In a shunt motor, if field winding is suddenly removed, how will the motor act?
8. Explain Fleming's left-hand rule.
9. If heavy load is kept during the starting of shunt motor, what will happen?
10. How do you find the stray losses in a shunt motor?

Brake test on DC compound motor

11. Draw the circuit for long and short shunt compound motor.
12. Mention the applications of DC compound motor.
13. What is cumulative and differential compound motor?
14. Can you reverse the direction of compound motor? How?
15. Plot the torque-speed characteristics for cumulative and differential compound motor.
16. What is meant by over and flat compounding?
17. In a compound motor, if series field is disconnected suddenly, how will the motor act?
18. Explain the procedure to conduct load test on compound motor.
19. If the series field winding is short circuited, how will it effect the speed of the motor?
20. Explain the effect of the losses in a DC machine.

Retardation test on DC shunt motor

21. How do you minimize the eddy current and hysteresis losses in a DC machine?
22. What are the factors that affect the skeleton of motor?
23. Define moment of inertia.
24. How do you reduce the inertia in DC motors?
25. Write the formula for hysteresis loss.
26. How the core losses show the effect on efficiency of the motor?
27. What is the effect of cross-magnetizing?
28. If AC supply is applied across the DC motor, what will happen?
29. If twice the rated voltage is applied across the armature of the DC motor, what will happen?
30. How do you increase the efficiency of the motor?

Swinburne's test on DC shunt machine

31. Mention the merits of Swinburne's test.
32. Why Swinburne's experiment should not be conducted on series motor?
33. The full load copper losses in shunt motor are 40 W. If the load is halved what would be the copper losses?

34. The full load core losses in shunt motor are 20 W. If the load is halved, what would be the core losses?
35. How do you control the speed of the DC motor?
36. Can the DC motor direction of rotation get reversed?
37. Explain the B-H curve of a material.
38. If both armature and filed windings are reversed in a DC motor, how the motor will work?
39. Mention the assumptions made in Swinburne's test.
40. If the shunt field crosses the 220 V DC supply, calculate the current flowing through the field having a resistance.

Hopkinson's test on two identical shunt machines

41. Mention the merits of Hopkinson's test.
42. Why Hopkinson's experiment should not be conducted on series motor?
43. The DC motor is running at 900 rpm and develops the back emf of 40 V. If the speed is dropped by 25% then the percentage of the speed reduces.
44. Mention the demerits of Hopkinson's test.
45. What is the other name for Hopkinson's test?
46. Explain the procedure to conduct Hopkinson's test.
47. Compare Hopkinson's test and Swinburne's test.
48. Why 3-point starter is used to start the DC motor?
49. Expand DPST.
50. How do you calculate the rating of the fuse?

Field test on two identical series machines

51. Field test is conducted on ____motors.
52. Why field test is conduced?
53. Explain the characteristics of series motor.
54. If the efficiency of the motor at full load is 70%, what is half-load efficiency?
55. Define torque.
56. How do you convert series motor into separately excited motor?
57. What is the condition for maximum efficiency of a DC motor?
58. Series motors produce high starting torques. Justify.
59. Mention the uses of DC series motor.
60. How do you control the speed of the DC series motor?

2.12 Objective questions

1. Constant losses do not depend on the changes in the
   (a) poles (b) load (c) current (d) none

2. The _____ winding is used in generators employed for welding.
   (a) lap (b) delta (c) simplex wave (d) duplex wave

3. If $E_b = V/2$, then the motor develop the_____power.
   (a) maximum (b) minimum (c) half (d) no effect

4. The voltage produced by the shunt generator is E when running at rated speed. What will be the generated voltage for twice the rated speed of the generator?
   (a) 4E (b) 2E (c) E/2 (d) E^2

5. If $E_b = 0$ suddenly in the motor then the motor
   (a) stop (b) continue to run (c) may burn (d) no effect

6. In a DC motor, unidirectional torque is produced with the help of
   (a) slip rings. (b) commutator (c) starter (d) thyristor control

7. The application of differentially compound motor is
   (a) frequent on–off cycle (b) low starting torque
   (c) high starting torque (d) variable speed

8. The function of dummy coil is to
   (a) reduce eddy current (b) reduce hysteresis loss
   (c) increase flux (d) mechanical balancing

9. The hysteresis loss is proportional to
   (a) frequency of magnetic reversals (b) maximum value of flux density
   (c) volume and grade of iron (d) all

10. The _____losses will increase the temperature of the machine.
    (a) constant (b) I^2R (c) stray (d) all

11. The purpose of lamination is to reduce_____losses.
    (a) stray (b) eddy current (c) I^2R (d) hysteresis

12. The core losses occur in
    (a) brushes (b) frame (c) winding (d) rotor core

13. The _____generator do not require residual magnetism to produce the voltage
    (a) compound (b) shunt (c) series (d) self-excited

14. A 110 V shunt generator running with zero field current at its rated speed. The voltage of the generator under open-circuited condition is
    (a) 50 V (b) about 2 V (c) 110 V (d) 75 V

15. The Hopkinson's experiment is performed on_____load.
    (a) zero (b) half (c) full (d) 3/4th

16. By interchanging______terminals the motion of the rotor of a DC series motor will be reversed
    (a) supply (b) field (c) load (d) none

Answers

1. b, 2. a, 3. c, 4. b, 5. c, 6. b, 7. b, 8. d, 9. d, 10. d, 11. b, 12. d, 13. d, 14. b, 15. c, 16. b

2.13 Exercise problems

1. A 220 V, 500 rpm and 50 A DC series motor. The total resistance of the machine is 0.25 Ω. Calculate the value of the additional resistance to be added in series with the motor such that the motor speed reduced to 300 rpm and the load torque should be half the previous value to the current.
2. A 220 V, 1,200 rpm DC shunt motor takes no load current of 10 A. The R_a and R_{sh} are 0.2 Ω and 250 Ω, respectively. Calculate the speed when the motor current is 50 A. Due to armature reaction, the field flux deteriorates by 4%.
3. A 500 V DC shunt motor takes 8 A at no load. Its armature resistance is 0.5 ohm and shunt field resistance is 250 ohm. Estimate the kW output and efficiency when the motor takes 40 A on full load.

3 Transformers

3.1 Introduction

The electromagnetic induction principles were introduced in 1831 by Faraday. It states that when alternatively varying flux links with a coil, an emf is induced in that coil and its magnitude is proportional to $\frac{d\phi}{dt}$, that is, rate of change of flux linkages. The voltage magnitude is also increased and decreased by the number of turns wounded in the coil. In electric power transmission, the transformer plays a key role in increasing and decreasing the voltage levels. The transformer rating is expressed in kVA as losses in a transformer do not depend upon the power factor.

3.2 Principle of operation

The transformer works on the principle of mutual inductance between two magnetically coupled coils. It consists of two coils and is linked by a magnetic flux, which is common for both the coils. The principle of working of a transformer is explained in Fig. 3.1.

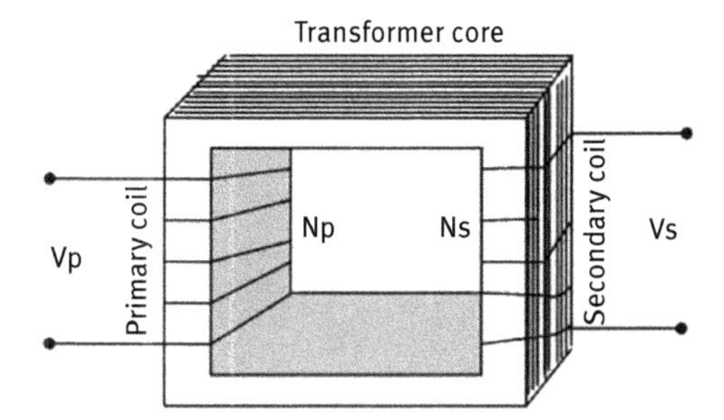

Fig. 3.1: Basic transformer construction.

The transformer consists of two coils, namely, primary and secondary windings. The coils are made up of copper or aluminum. But usually transformer winding is made up of copper only. The two windings are wounded on laminated core and it reduces reluctance of the path of the magnetic flux. The core laminations decrease the losses due to eddy current and they are made up of silicon sheet steel laminations that reduce the hysteresis losses.

The working principle of transformer is explained as follows:

When an AC voltage is connected to the primary winding or coil, it draws the exciting current, which establishes the magnetomotive force (mmf) in the primary

https://doi.org/10.1515/9783110682274-003

coil and this mmf produces flux that varies with respect to time. According to Faraday's laws of electromagnetic induction, if a changing flux links with a coil, an emf is induced in that coil. This emf is called self-induced emf. The direction of the flux will be identified by the thumb rule and the same flux will flow through the magnetic core and link the secondary coil; hence, emf is induced in the secondary coil. This emf is called mutually induced emf if the secondary circuit is closed current, which starts flowing through the load. The magnitude of induced emf is determined by the Faraday's second law, that is, $E = N\frac{d\phi}{dt}$.

3.3 Types of transformers

The transformers are categorized based on the winding provided on the core. According to the design, the transformers are of two types, namely core- and shell-type transformer.

(i) Core-type transformer

In this type, the core is surrounded by the winding or coil. The coils are wounded in cylindrical shape around the core. The coils are insulated with different insulating materials such as paper, cloth, varnish and mica sheets. The core-type transformer construction is shown in Fig. 3.2.

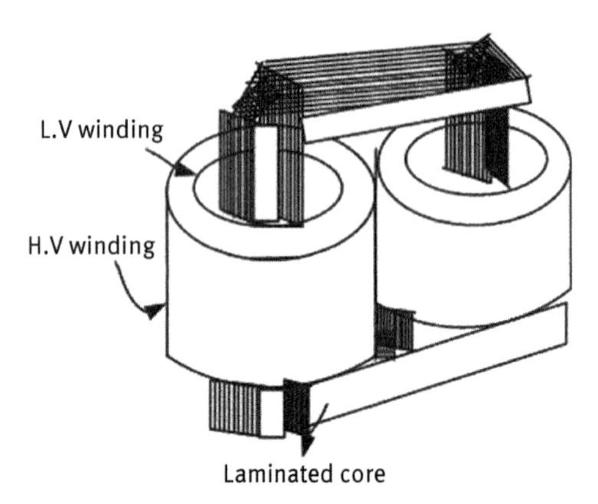

Fig. 3.2: Transformer core laminations.

The LV windings are wounded near to the core and HV windings are placed above the LV winding.

(ii) Shell-type transformer

The shell-type transformer construction is shown in Fig. 3.3. In this type of transformers, the core is enclosed with a significant portion of the winding.

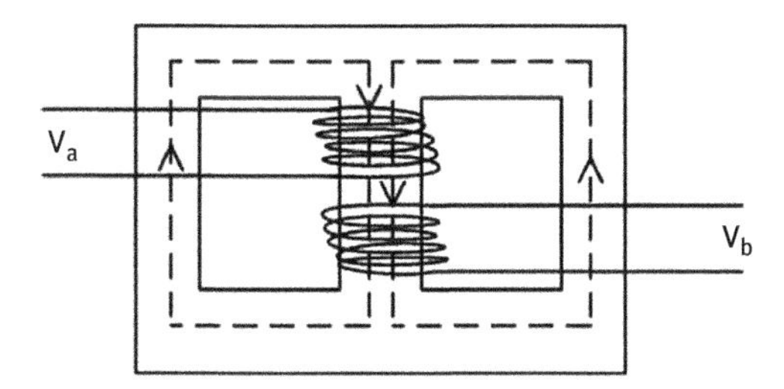

Fig. 3.3: Shell-type transformer.

3.4 Voltage equation of a transformer

Let us consider that the magnetic flux varies sinusoidally, $\phi = \phi_{max} \sin \omega t$,

$$\text{The emf induced in the winding, } E = N\frac{d\phi}{dt} \tag{3.1}$$

$$N\frac{d(\phi_{max} \sin \omega t)}{dt} = N\omega\phi_{max} \cos \omega t$$

$$\Rightarrow E_{max} = N\,\omega\,\phi_{max}$$

$$E_{rms} = \frac{\phi_{max}}{\sqrt{2}}\,N\omega = \frac{\phi_{max}}{\sqrt{2}}\,N \times 2\pi f = 4.44\,\phi_{max} Nf \tag{3.2}$$

Equation (3.2) is known as the voltage equation of the transformer.

If N_P and N_s are the number of primary and secondary turns, then the emf induced in the primary winding = $E_p = E_1 = 4.44\,\phi_{max}\,N_p\,f$

The emf induced in the secondary winding = $E_s = E_2 = 4.44\,\phi_{max}\,N_s\,f$

$$\frac{E_P}{E_S} = \frac{N_P}{N_S} \text{ or } \frac{E_1}{E_2} = \frac{N_1}{N_2}$$

Turn ratio

The ratio between the number of turns in the primary winding to turns in the secondary winding is called as turns ratio, that is, $= \frac{Np}{Ns}$.

Transformation ratio

The ratio between the emf induced in the secondary winding to emf induced in the primary winding is called transformation ratio. It is represented by K.

$$\frac{E_S}{E_P} = \frac{N_S}{N_P} = K \tag{3.3}$$

3.5 Equivalent circuit of the transformer

The equivalent circuit of a transformer is essential in finding the behavior and to analyze the performance. The equivalent circuit of the transformer is drawn viewing from the primary side and also drawn viewing from the secondary side. Let us draw the equivalent circuit of transformer. The transformer's primary resistance R_1 and reactance X_1 are in series and the excitation branch parameters R_c and X_O are considered parallel and similarly the secondary winding resistance R_2 and reactance X_2 are in series. The equivalent circuit of transformer is shown in Fig. 3.5. The phasor diagram is shown in Fig. 3.4.

3.5.1 Transformer equivalent circuit referred to primary side

Let us consider the transformation ratio to be $\frac{E_S}{E_P} = \frac{N_S}{N_P} = K$.

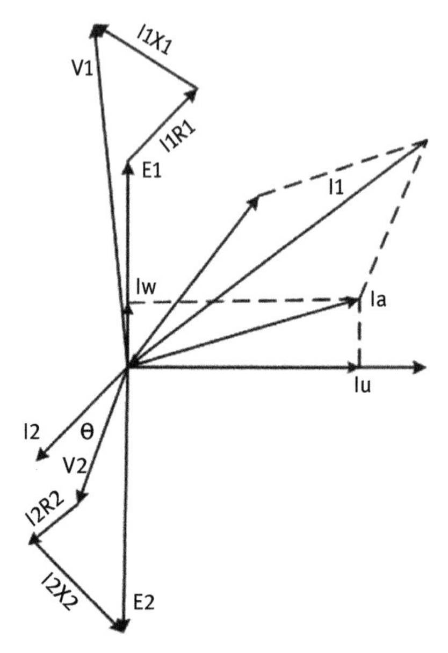

Fig. 3.4: Transformer vector diagram on load.

V_1 is the voltage supplied to the primary winding;
E_1 is the induced voltage in the primary winding;
I_1 is the current drawn by the transformer;
Z_1 is the impedance of the primary winding = $R_1 + jX_1$.

The voltage drop in the primary winding due to current I_1 is I_1Z_1. Therefore, the voltage equation can be written as

$$E_1 = V_1 - I_1Z_1 = V_1 - (I_1R_1 + jI_1X_1) \tag{3.4}$$

Similarly, the voltage equation of the secondary can be written as

$$E_2 = V_2 - I_2Z_2 = V_2 - (I_2R_2 + jI_2X_2)$$

The current of secondary is I_2. So, the voltage E_2 across the secondary winding is partly dropped by $I_2 Z_2$ or $I_2 R_2 + jI_2 X_2$ before it appears across the load.

This parallel path of current is known as excitation branch of an equivalent circuit of the transformer.

The excitation branch resistance $R_0 = \frac{E_1}{I_w}$

The excitation branch reactance $X_0 = \frac{E_1}{I_\mu}$

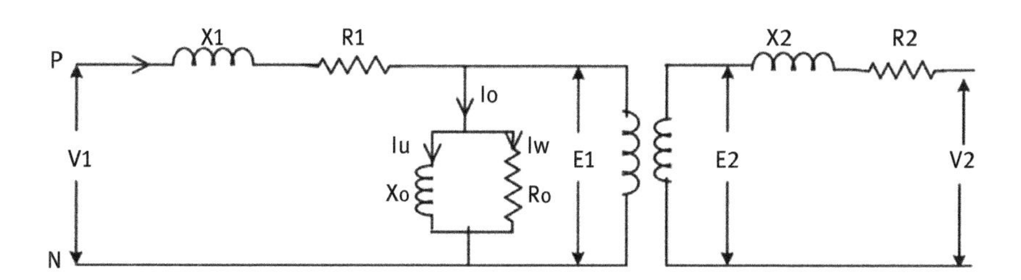

Fig. 3.5: Complete equivalent circuit of transformer.

The load component I_2' flows through the primary winding of the transformer and induced voltage across the winding is E_1 as shown in Fig. 3.5. This induced voltage E_1 transforms to secondary and it is E_2 and load component of primary current I_2' is transformed to secondary as secondary current I_2. The current of secondary is I_2. So, the voltage E_2 across secondary winding is partly dropped by I_2Z_2 or $I_2R_2 + jI_2X_2$ before it appears across the load. The load voltage is V_2.

The complete equivalent circuit referred to primary is shown in Fig. 3.6.

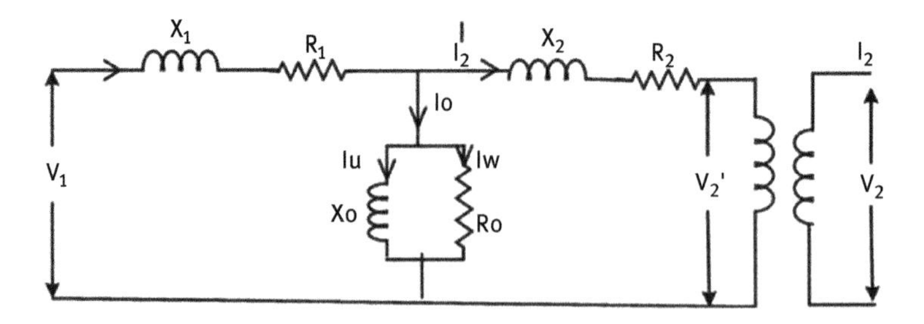

Fig. 3.6: Equivalent circuit of transformer referred to primary.

From the principle of transformer: input VA = output VA

$$I_1 N_1 = I_2 N_2$$

$$I_1 = I_2 \frac{N_2}{N_1} = \frac{I_2}{K}$$

$$I_2{}' \times N_1 = I_2 \times N_2 \tag{3.5}$$

$$I_2 = I_2{}' \frac{N_1}{N_2} = K I_2{}'$$

Hence, the secondary impedance of transformer referred to primary is written from the above equation:

$$Z_2{}' = K^2 Z_2 \tag{3.6}$$

$$R_2{}' = K^2 R_2 \text{ and } X_2{}' = K^2 X_2$$

Similarly, equivalent circuit of the transformer referred to secondary is drawn.

3.6 Losses, efficiency and regulation of a transformer

3.6.1 Losses

While transforming electrical power from primary winding to secondary winding, the entire electrical power is not transformed. Hence, a little portion of the input power is wasted in core and winding of the transformer known as loss of power or loss. The major drawbacks of these losses are temperature rise, reduction in efficiency and reduction in lifetime of the transformer.

These losses are of two types: (i) constant losses and (ii) variable losses.

These are discussed in Section 2.7, page 45.

3.6.2 Efficiency of transformer

The efficiency of a transformer is defined as the ratio between the output power to the input power:

$$\text{Efficiency}(\eta)=\frac{\text{output power}}{\text{input power}}=\frac{\text{input power}-\text{losses}}{\text{input power}}=\frac{\text{output power}}{\text{output power}+\text{losses}} \tag{3.7}$$

$$\text{Efficiency}=\frac{\text{input power}-\text{losses}}{\text{input power}}=1-\frac{\text{losses}}{\text{input power}}=1-\frac{I_1^2R_1+W_i}{\text{input power}}$$

$$\text{Efficiency},\eta=1-\frac{I_1^2R_1+W_i}{V_1I_1\text{Cos}\phi}\ \text{or}$$

$$\text{Efficiency}\ (\eta)=\frac{\text{output power}}{\text{output power}+\text{losses}}=\frac{V_2I_2\text{Cos}\phi}{V_2I_2\text{Cos}\phi+I_1^2R_1+W_i} \tag{3.8}$$

$$\text{Condition for maximum efficiency},\eta=\frac{d(\text{efficiency})}{d\,I_1}=0$$

$$\frac{R_1}{V_1\ \text{Cos}\phi}=\frac{W_i}{V_1\ I_1^2\text{Cos}\phi}$$

Hence, the efficiency is maximum $W_i = I_1^2R_1$.

It reveals from the above equation if the transformer copper loss are equal to iron losses, then the transformer exhibits the maximum efficiency.

3.6.3 Voltage regulation of transformer

In a transformer, the amount of the voltage induced in secondary winding should be equal to the voltage available across the terminal of the load. But due to resistance of the winding, it may not be attained. Therefore, it is required to calculate the amount of voltage drop in the winding from no load to full load.

The difference in the voltage from no load to full load expressed as a percentage of the no-load voltage is known as voltage regulation of the transformer. It is commonly used in power engineering to describe the voltage drop in transformer and transmission lines:

$$\text{Percentage of voltage regulation}=\frac{E_2-V_2}{E_2}\times 100$$

The difference in no-load and full-load voltage E_2-V_2 is not known directly. Hence, equation is derived for voltage regulation with the phasor diagram shown in Fig. 3.7.

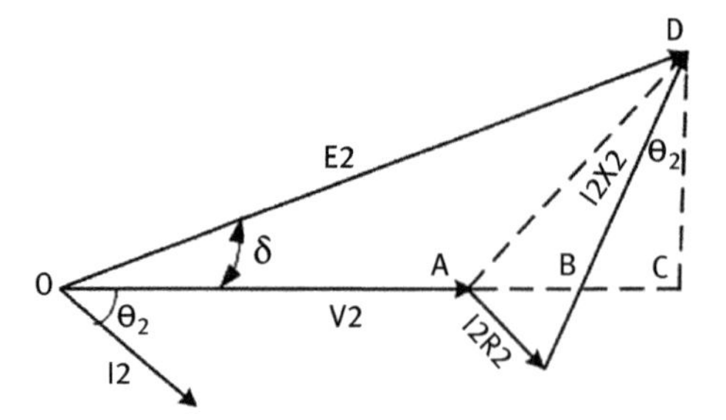

Fig. 3.7: Phasor diagrams of transformer for lagging pf.

The transformer phasor diagram is drawn for lagging power factor load. It is shown in Fig. 3.7.

Let

V_2 be the secondary terminal voltage;

E_2 the no-load voltage of the secondary winding;

I_2 the load current drawn by the transformer;

Z_2 the impedance of the secondary winding = $R_2 + jX_2$;

Cos θ_2 the power factor of the load.

From the phasor,

$$\overline{OC} = \overline{OA} + \overline{AB} + \overline{BC}$$

Here $\overline{OA} = \overline{V_2}$

$$AB = AE\ \text{Cos}\,\theta_2 = I_2R_2\ \text{Cos}\,\theta_2$$

$$BC = DE\ \text{Sin}\,\theta_2 = I_2X_2\ \text{Sin}\,\theta_2$$

The angle between OC and OD may be very small, so it can be neglected and OD is considered nearly equal to OC, that is,

$$\overline{E_2} = \overline{OC} = \overline{OA} + \overline{AB} + \overline{BC}$$

$$\overline{E_2} = \overline{OC} = V_2 + I_2R_2\text{Cos}\,\theta_2 + I_2X_2\text{Sin}\,\theta_2$$

Voltage regulation of transformer at lagging power factor,

$$= \frac{E_2 - V_2}{E_2} \times 100 = \frac{I_2\,R_2\cos\theta_2 + I_2\,X_2\sin\theta_2}{E_2} \times 100 \qquad (3.9)$$

Similarly, voltage regulation of transformer at a leading power factor,

$$\frac{I_2\,R_2\cos\theta_2 - I_2\,X_2\sin\theta_2}{E_2} \times 100$$

3.7 Testing of transformers

3.7.1 Open-circuit and short-circuit tests on single-phase transformer

The efficiency and percentage regulation of a single-phase transformer is evaluated by the open-circuit (OC) and short-circuit (SC) tests.

3.7.1.1 Apparatus required

Table 3.1: List of required apparatus.

Name	Range	Quantity
Single-phase transformer	115/230 V, 3 kVA	1
Single-phase autotransformer	230/0–270, 8 A	1
Ammeter	0–2.5/5 A, MI	1
Ammeter	0–15 A, MI	1
Voltmeter	0–30/60 V, MI	1
Voltmeter	0–150/300 V, MI	1
Wattmeter	LPF, 75/150/300 V, and 2.5/5 A	1
Wattmeter	UPF, 75/150/300 V and 5/10 A	1

3.7.1.2 Circuit diagram

OC test

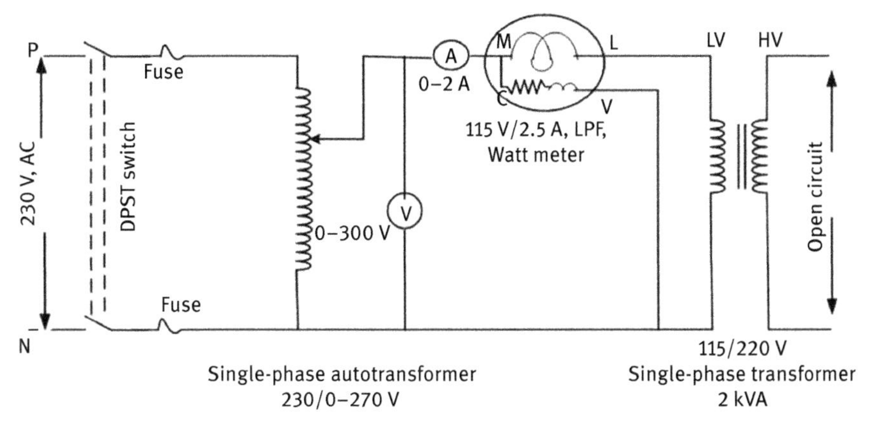

Fig. 3.8: Circuit diagram of single-phase transformer: open-circuit test.

SC test

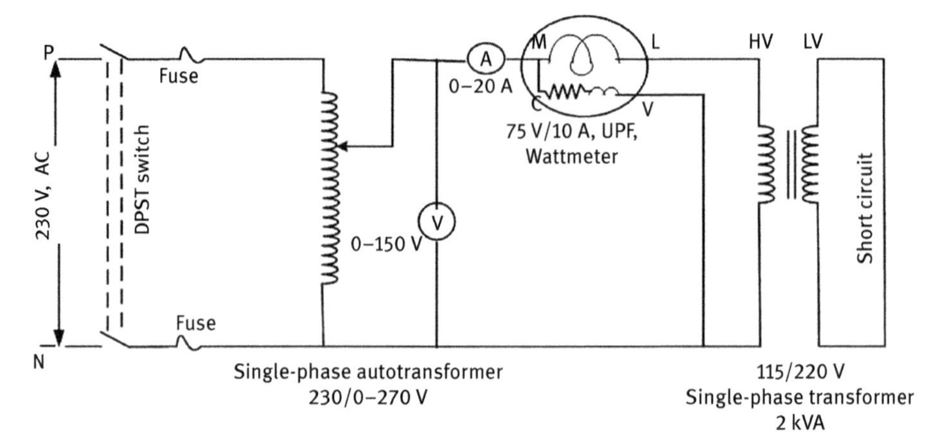

Fig. 3.9: Circuit diagram of single-phase transformer: short-circuit test.

3.7.1.3 Theory

OC or no-load test

The OC test is used to find out the core losses and excitation branch parameters (shunt branch). This test is conducted on LV side, and HV side is kept open. During the OC test, the transformer draws exciting current (I_0) to about 2–5% of the rated current. The primary resistance and reactance (R_0 and X_0) are very small; hence, the applied voltage $V_1 = E_1$. Therefore, the power drawn by the transformer under no-load condition will meet the core losses only.

Let
V_1 be the supply voltage;
I_0 the no-load current;
P_0 the no-load power input;

$$P_0 = V_1 I_0 \cos \phi_0; \ \cos \phi_0 = \frac{P_0}{V_1 I_0};$$

$$I_w = I_0 \cos \phi_0, \ I_\mu = I_0 \sin \phi_0;$$

$$R_0 = \frac{V_1}{I_w}, \qquad X_0 = \frac{V_1}{I_\mu}.$$

SC test

The SC test is used to find out the copper losses and primary and secondary winding parameters. This test is conducted on HV side, and LV side is short-circuited. Since small voltage around 4–8% of the rated voltage is needed to circulate full-load current of the transformer, the exciting current drawn under full-load condition is around 0.2–0.4%. Hence, the shunt branch parameters are neglected. Therefore, the power drawn by the transformer under full-load condition will meet the copper losses only. The core losses are negligible since very less voltage is applied to circulate full-load current.

Let

V_{sc} be the SC voltage;

I_{sc} the full-load current;

P_{sc} the full-load power input;

$$Z_{01} = \frac{V_{sc}}{I_{sc}} = \sqrt{R_{01}^2 + X_{01}^2};$$

$$\text{Equivalent resistance, } R_{01} = \frac{Psc}{Isc^2};$$

$$\text{Equivalent reactance, } X_{01} = \sqrt{Z_{01}^2 - R_{01}^2};$$

$$\%\,\text{Regulation} = \frac{I_1R_1\cos\theta_1 - I_1X_1\sin\theta_1}{V_1} \times 100;$$

$$\%\,\text{Efficiency} = \frac{\text{kVA}\cos\theta}{\text{kVA}\cos\theta + \text{losses}};$$

3.7.1.4 Procedure

OC test

1. The details of the transformer should be noted.
2. The connections of various terminals are connected as per the circuit with Double Pole Single Throw (DPST) off position.
3. Make sure that all measuring instruments are connected to the LV side of the transformer, and HV side must be open-circuited.
4. Using single-phase autotransformer, gradually apply the rated primary voltage to the LV winding (single-phase supply of 230 V).
5. Record all the meter readings and bring back to the 0 V position of the autotransformer. Switch off the panel.
6. Calculate the necessary parameters.

SC test

1. The connections of various terminals are connected as per the circuit with DPST off position.
2. Make sure that all measuring instruments are connected to the HV side of the transformer, and LV side must be short-circuited.
3. Using single-phase autotransformer, gradually apply the voltage to the HV winding (about 30–40 V).

4. Record all the meter readings and bring back to the 0 V position of the autotransformer. Switch off the panel.
5. Calculate the necessary parameters.

3.7.1.5 Observation table

For OC test

Table 3.2: Observation table.

S. no.	Voltmeter reading (V_0)	Ammeter reading (I_0)	Wattmeter reading (W_0)	R_{01} (Ω)	X_{01} (Ω)	Cos ϕ_0
1						

For SC test

Table 3.3: Observation table.

S. no.	Voltmeter reading (V_{SC})	Ammeter reading (I_{SC})	Wattmeter reading (W_{SC})	R_{02} (Ω)	X_{02} (Ω)
1					

3.7.1.6 Model calculations

From the tabulated data, the parameters of shunt branch (R_O, X_O) and primary, secondary winding parameters (R_{01}, R_{02}, X_{01} and X_{02}) are calculated:

$$R_O = V_1/I_w,\ X_O = V_1/I_\mu, \text{where } I_w = I_0 \cos \phi_O,\ I_\mu = I_O \sin \phi_O$$

$$\text{Equivalent resistance, } R_{01} = P_{SC}/(I_{SC})^2$$

$$\text{Equivalent reactance, } X_{01} = \sqrt{Z_{01}^2 - R_{01}^2}$$

At half the rated load

P_{SC} is the full-load copper losses;
P_O the core losses (these losses are constant);
cos ϕ the power factor.

Calculate the copper losses at half load.

$$\text{At half load, the copper losses are} = P_{SC} \times \left(\frac{1}{2}\right)^2 \text{ watts}$$

$$\text{The output power at half load} = \frac{1}{2}(\text{kVA}) \times \cos\phi$$

Input power = output power + copper losses + constant losses

$$\%\,\text{Efficiency} = \frac{\text{output power}}{\text{input power}} \times 100 = \frac{\frac{1}{2}\text{kVA}\cos\theta}{\frac{1}{2}\text{kVA}\cos\theta + \text{losses}} \times 100$$

$$\%\,\text{Regulation} = \frac{I_1R_1\cos\theta_1 - I_1X_1\sin\theta_1}{V_1} \times 100$$ "+" for lagging and "−" for leading power factors.

Table 3.4: Observation table.

S. no.	Power factor	% Regulation	
		Lag	Lead
1	1		
2	0.8		
3	0.6		
4	0.4		
5	0.2		

3.7.1.7 Model graph

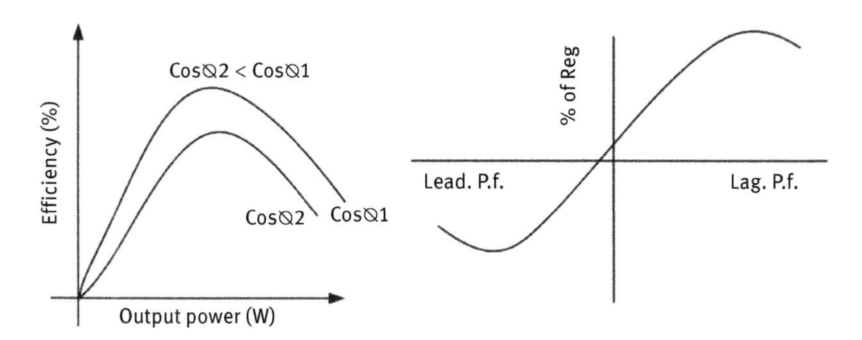

Fig. 3.10: Model graphs of transformer.

3.7.1.8 Practical calculations

The OC and SC tests are conducted in the laboratory with the following ratings of the transformer.

Rating of the transformer : 3 kVA
Rated voltage : 230 V
Full-load current : 13 A

For OC test

Table 3.5: Practical observations.

S. no.	Voltmeter reading (V_0)	Ammeter reading (I_0)	Wattmeter reading (W_0)	R_{01} (Ω)	X_{01} (Ω)	$\cos\phi_0$
1	115	1.7	88	150.32	76.15	0.45

For SC test

Table 3.6: Practical observations.

S. no.	Voltmeter reading (V_{SC})	Ammeter reading (I_{SC})	Wattmeter reading (W_{SC})	R_{02} (Ω)	X_{02} (Ω)
1	21	13.04	240	0.705	1.447

3.7.1.9 Calculations

OC test

No-load power factor $\cos\phi_0 = \frac{W_0}{V_0 I_0} = \frac{88}{115 \times 1.75} = 0.45$

Working component of no-load current $I_W = I_0 \cos\phi_0 = 1.7 \times 0.45 = 0.765$ A

$$\phi_0 = \cos^{-1}(0.45) = 63.25^0$$

Magnetizing component of no-load current $I_\mu = I_0 \sin\phi_0 = 1.7 \times \sin(63.25) = 1.51$ A

Equivalent circuit parameters $R_{01} = \frac{V_1}{I_W} = \frac{115}{0.765} = 150.32\ \Omega$

$$X_{01} = \frac{V_1}{I_\mu} = \frac{115}{1.51} = 76.15\ \Omega$$

SC test

Total resistance, impedance and reactance are referred to primary

$$R_{02} = \frac{W_{SC}}{I_{SC} \times I_{SC}} = \frac{120}{13.04 \times 13.04} = 0.705\ \Omega$$

$$Z_{02} = \frac{V_{SC}}{I_{SC}} = \frac{21}{13.04} = 1.61\ \Omega$$

$$X_{02} = \sqrt{(Z_{02})^2 - (R_{02})^2} = \sqrt{(1.61)^2 - (0.705)^2} = 1.447\ \Omega$$

$$\text{Percentage regulation} = \frac{I_2 R_{02} \cos\phi \pm I_2 X_{02} \sin\phi}{V_2}$$

1. When $\cos\phi = 1, \quad \sin\phi = 0$

$$\%\,\text{Regulation} = \frac{13.04 \times 0.75 \times 1 + 13.04 \times 1.447 \times 0}{230} \times 100 = 4.65\%$$

For leading: 4.25

2. When $\cos\phi = 0.8, \quad \sin\phi = 0.6$

$$\%\,\text{Regulation} = \frac{13.04 \times 0.705 \times 0.8 + 13.04 \times 1.447 \times 0.6}{230} \times 100 = 8.32\%$$

For leading: −1.52

3. When $\cos\phi = 0.6\,\text{lag}, \quad \sin\phi = 0.8$

$$\%\,\text{Regulation} = \frac{13.04 \times 0.75 \times 0.6 + 13.04 \times 1.447 \times 0.8}{230} \times 100 = 8.95\%$$

For leading: −4.16

4. When $\cos\phi = 0.4\,\text{lag}, \quad \sin\phi = 0.91$

$$\%\,\text{Regulation} = \frac{13.04 \times 0.75 \times 0.4 + 13.04 \times 1.447 \times 0.91}{230} \times 100 = 9.16\%$$

For leading: −5.76

Efficiency

$$\%\,\text{Efficiency}\ (\eta) = \frac{\text{kVA} \times \cos\phi}{\text{kVA}\cos\phi + \text{losses}}$$

For cos ϕ = 1

i) $\%\eta$ at full load $= \dfrac{\text{kVA} \times \cos\phi}{\text{kVA}\cos\phi + \text{losses}} \times 100 = \dfrac{3 \times 10^3}{3 \times 10^3 + 22 + 220} \times 100 = 92.53$

ii) $\%\eta$ at half load $= \dfrac{(1/2)\text{kVA} \times \cos\phi}{(1/2)\text{kVA}\cos\phi + (1/4)\ \text{losses}} \times 100$

$$= \frac{(1/2) 3 \times 10^3}{(1/2) 3 \times 10^3 + 22 + (1/4) 220} \times 100 = 94.816$$

iii) $\%\eta$ at $\frac{3}{4}$ load

$$= \frac{(3/4)\text{kVA} \times \cos\phi}{(3/4)\text{kVA} \cos\phi + (3/4)^2 \text{ losses}} \times 100$$

$$= \frac{(3/4)3 \times 10^3}{(3/4)3 \times 10^3 + 22 + (3/4)^2 220} \times 100 = 96.47$$

iv) $\%\eta$ at $\frac{1}{4}$ load

$$= \frac{(1/4)\text{kVA} \times \cos\phi}{(1/4)\text{kVA} \cos\phi + (1/4)^2 \text{ losses}} \times 100$$

$$= \frac{(1/4)3 \times 10^3}{(1/4)3 \times 10^3 + (1/4)^2 220} \times 100 = 95.29$$

The calculated values are tabulated.

Table 3.7: Practical observations.

S. No.	Load	Efficiency
1	Full load	92.53
2	3/4	96.47
3	1/2	94.816
4	1/4	95.29

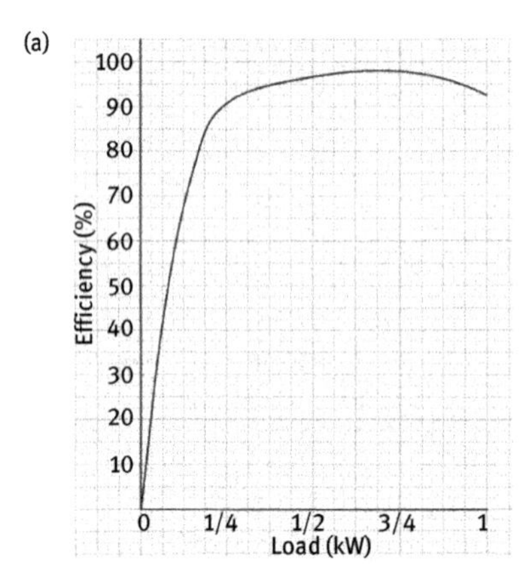

Fig. 3.11a: Efficiency versus load curve of single-phase transformer.

(b)

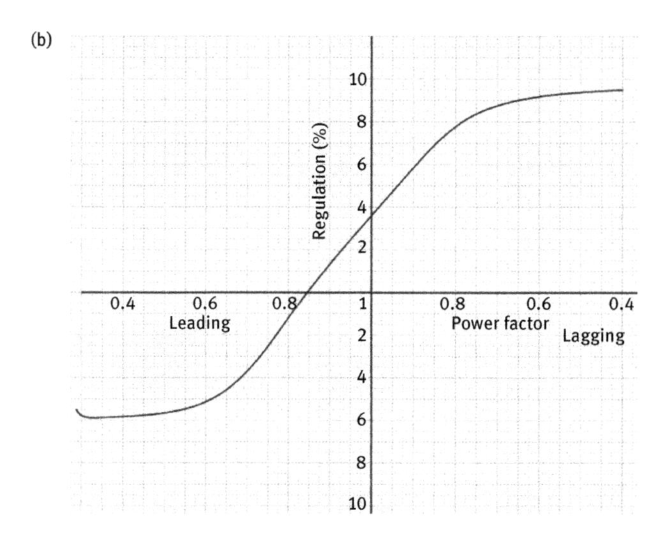

Fig. 3.11b: Power factor versus % regulation curve of single-phase transformer.

Power factor versus regulation

Table 3.8: Practical observations.

S. no.	Power factor	% Regulation	
		Lag	**Lead**
1	1	4.25	4.25
2	0.8	8.32	−1.52
3	0.6	8.95	−4.16
4	0.4	9.16	−5.76

3.7.1.10 Result

Perform OC and SC tests and calculate the efficiency and regulation of the transformer at various power factors.

3.7.2 Separation of core losses of a single-phase transformer

The performance of transformer depends on the quality of the core of a transformer. Therefore, the transformer possesses low core losses. Hence, it is required to calculate core losses of a transformer and the portion of eddy current losses, and hysteresis losses also to be calculated.

3.7.2.1 Apparatus required

Table 3.9: List of required apparatus.

Name	Range	Quantity
Voltmeter	0–250 V (MI)	1 no.
Ammeter	0–3 A (MI)	1 no.
Wattmeter	0–10/20 A, 150/300/600 V, LPF	2 no.
Rheostat	185 Ω/2.3 A	2 no.
Rheostat	100 Ω/5 A	1 no.
Rheostat	400 Ω/1.7 A	1 no.

3.7.2.2 Circuit diagram

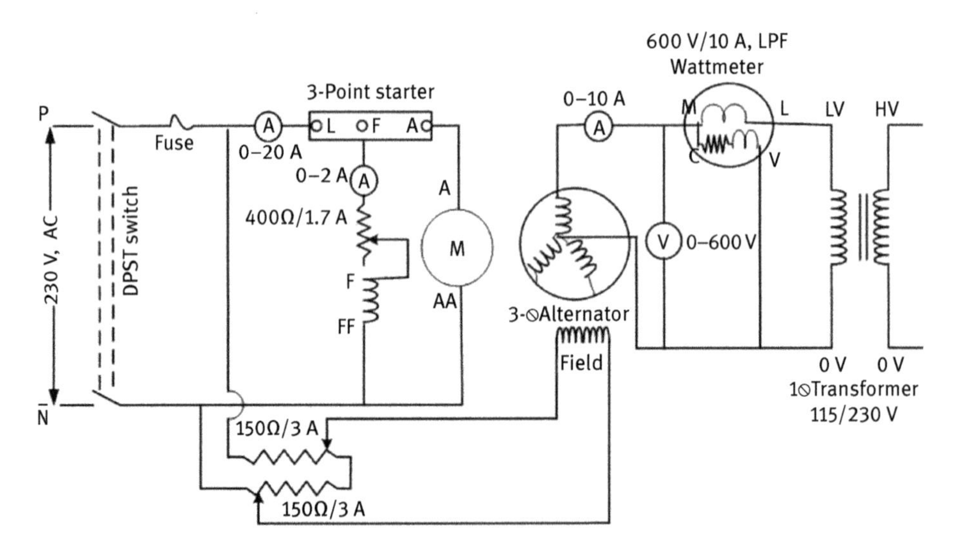

Fig. 3.12: Circuit diagram single-phase transformer: separation of core losses test.

3.7.2.3 Theory

The losses that occur in a transformer are known as core losses. These losses are very useful in estimating the quality of the core. These losses are also termed as constant losses as these do not depend on the load. The core losses are categorized into two types, namely hysteresis loss and eddy current losses. These are described in Section 2.7.

The hysteresis loss is calculated using the Steinmetz formula:

$$\text{Hysteresis loss, } P_h = \eta B_{max}^{1.6} fV$$

$$\text{Eddy current loss, } P_e = \eta B_{max}^{2} f^2 t^2$$

where η is the Steinmetz coefficient of hysteresis;
B_{max} is the maximum flux density;
f the frequency;
V the volume of the core in m^3;
t the thickness of the lamination in cm.
The iron loss will be expressed by $W_i = Af + Bf^2$

$$W_i/f = A + Bf$$

3.7.2.4 Procedure

1. The details of the transformer should be noted.
2. The connections of various terminals are connected as per the circuit with DPST off position.
3. Keep the motor field rheostat (R_{fm}) in the minimum position, armature rheostat in maximum position and start the motor by closing the switch and operating the starter slowly. Bring the motor speed up to the rated. Make sure that the generator field current should be zero initially.
4. Now gradually increase the generator field till the rated voltage is generated at rated frequency (50 Hz) and record the meter readings.
5. Now calculate $\frac{V}{f}$ ratio for step 4 and maintain the same ratio throughout the experiment.
6. Now by varying the V and f and maintain the $\frac{V}{f}$ ratio same as step 5, record meter readings.
7. Repeat step 6 and take the readings in meters.
8. Draw the plot between core loss per frequency (W_0/f) versus frequency "f" and calculate the coefficients A and B.

3.7.2.5 Observation table

Table 3.10: Observation table.

S. no.	Speed of the alternator (rpm)	Supply frequency, f (Hz)	Primary voltage, V (V)	Wattmeter readings, Wo (W)	Iron or core loss, W_i (W)	W_i/f
1						
2						
3						
4						

3.7.2.6 Calculations

1. Frequency $(f) = \frac{PN}{120}$
2. Hysteresis loss $(W_h) = A * f$
3. Eddy current loss $(W_e) = B * f^2$
4. Iron loss or core loss $(W_i) = W_e + W_h$

3.7.2.7 Model graph

The plot is drawn between frequency versus W_i/f.

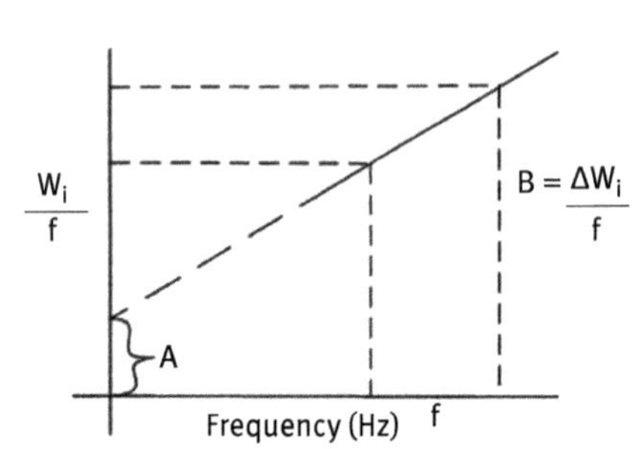

Fig. 3.13: Model graph of frequency versus (W_i/f) curve.

3.7.2.8 Practical calculations

The separation of losses experiment is performed on the following rating of the single-phase transformer and readings are tabulated.

Table 3.11: Practical observations.

Motor	Generator
Rating = 3.5 kW, 1,500 rpm	Output power = 5 kVA
Current = 27.8 A	Current = 7 A

3.7.2.9 Observation table

Table 3.12: Practical observations.

S. No.	Speed of the prime mover N, (rpm)	Supply frequency, f(Hz)	Primary voltage, V (V)	V/f ratio	Iron or core loss W_i (W)	W_i/f
1	1,500	50	230	4.6	15	0.3
2	1,200	40.2	185	4.6	10.4	0.26
3	912	30.4	140	4.6	7.08	0.233

From the graph at 40.2 Hz,
A = 0.0523, B = 0.0049.
Total core losses $W_p = A_f + B_f{}^2$

At f = 50 Hz
Hysteresis loss $P_h = Af = 0.0523 \times 50 = 2.615$
Eddy current loss $P_e = Bf^2 = 0.0049 \times 50^2 = 12.25$
Core loss $W_i = P_h + P_e = 2.615 + 12.25 = 14.865$ watts

At f = 40.2 Hz
$P_h = Af = 0.0523 \times 40.2 = 2.10$
$P_e = Bf^2 = 0.0049 \times 40.2 = 7.9$
$W_i = P_h + P_e = 2.10 + 7.9 = 10.04$

3.7.2.10 Result

The eddy current losses and hysteresis losses are separated from the core losses of 2 kVA single-phase transformer.

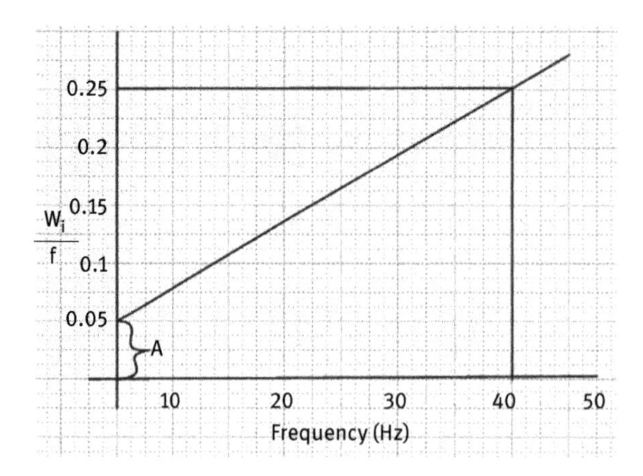

Fig. 3.14: Graph of frequency versus (Wi/f) curve.

3.7.3 Sumpner's test

The Sumpner's test is also called back-to-back test. In this test, the efficiency and losses of a transformer are at rated load conditions.

3.7.3.1 Apparatus required

Table 3.13: List of required apparatus.

Name	Range	Quantity
Voltmeter	0–300 V (MI)	1
Voltmeter	0–600 V (MI)	1
Ammeter	0–2 A (MI)	1
Ammeter	0–10 A (MI)	1
Wattmeter	0–600 V/0–20 A, UPF	1
Wattmeter	0–300 V/0–5 A, LPF	1
Connecting wires		

3.7.3.2 Circuit diagram

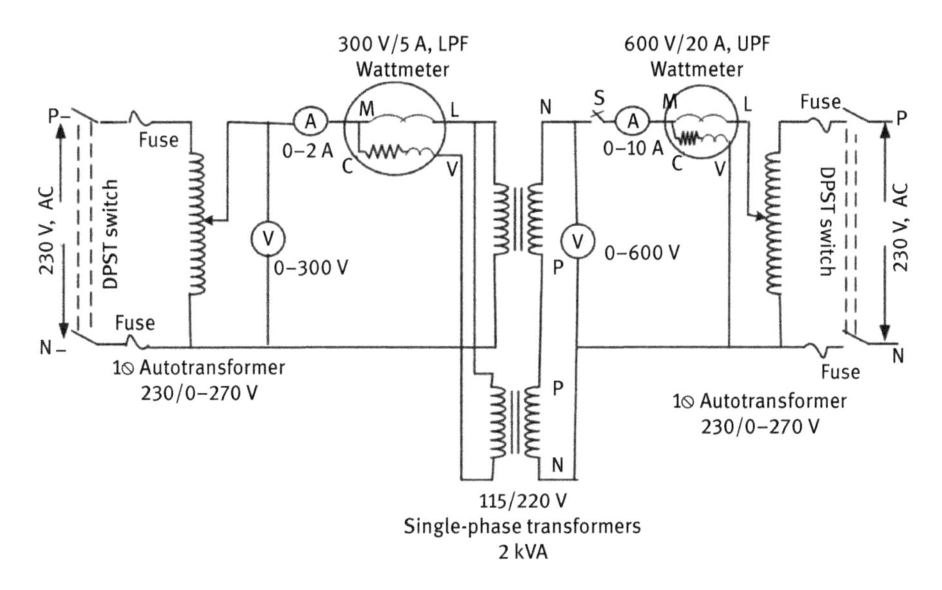

Fig. 3.15: Circuit diagram of single-phase transformer: Sumpner's test.

3.7.3.3 Theory

The SC test is conducted with actual load and is convenient to put actual load in small transformer only as the full-load current is very small. But in large size transformer, the actual loading involves huge power loss. This drawback is overcome in Sumpner's test.

This test is conducted without the actual load connected on the transformer. This test is conducted on two identical transformers. The primary windings are connected in parallel and secondary windings of the transformer are connected in series but in phase opposition.

If the secondary winding of the transformer is open-circuited and primaries are connected to the rated supply voltage, as the secondary winding of the transformer are open-circuited, no current will flow. Hence, the transformers will draw only exciting current, that is, $2I_o$. The power drawn from the source is $2P_i$, i.e. to meet both the transformer core losses. Now to circulate full-load current in the secondary of the transformer requires small voltage. This current will circulate in the secondary of the transformer only. Hence, the wattmeter connected in the secondary side will read the power loss due to the winding resistance, that is, copper losses. The total copper losses are $2P_c$.

The total core losses = primary side wattmeter reading = P_0

The core loss of each transformer = $\frac{P_i}{2}$

The total copper losses = secondary side wattmeter reading = P_{sc}

The copper loss of each transformer $= \frac{P_{SC}}{2}$

$$\text{Efficiency} = \frac{\text{kVA}\cos\theta}{\text{kVA}\cos\theta + \text{losses}} \times 100$$

Losses in each transformer $= \frac{P_i}{2} + \frac{P_{SC}}{2}$
Efficiency of each transformer $= \frac{\text{kVA}\cos\theta}{\text{kVA}\cos\theta + \frac{P_i}{2} + \frac{P_{SC}}{2}}$

3.7.3.4 Procedure

1. The details of the transformer should be noted.
2. The connections of various terminals are connected as per the circuit with DPST off position. Make sure that switch "S" is open.
3. Now close the DPST in the primary side of the transformer and using single-phase autotransformer, gradually apply the rated voltage of the transformer.
4. Now check the voltmeter reading connected across the switch. It should read "Zero," if it reads double the supply voltage, then interchange the connections of the secondary windings of the transformer.
5. If the voltmeter reading connected across the switch reads "Zero," then close the switch "S."
6. Now close the DPST of the secondary of the transformer and apply rated current is circulating through the secondary winding of the transformer.
7. Note down the readings.

3.7.3.5 Observation table

On the primary side:

Table 3.14: Practical observations.

S. no.	Voltage of primary side, V_1 (V)	Current of primary side, I_0 (A)	Power on primary side, P_0 (W)
1	230	4.6	420

On the secondary side:

Table 3.15: Practical observations.

S. no.	Voltage of secondary side, V_{SC} (V)	Current of secondary side, I_{SC} (A)	Power on secondary side, P_{SC} (W)
1	60	5	50

3.7.3.6 Model graph

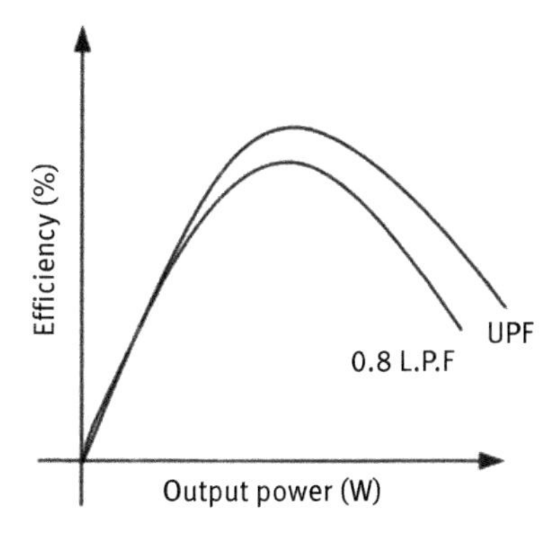

Fig. 3.16: Model graph of output power versus efficiency.

3.7.3.7 Practical calculations

The Sumpner's test is conducted in the laboratory with the following ratings of the transformer and record the data.

Rating of the transformer : 3 kVA
Rated voltage : 230 V
Full-load current : 13 A

On the primary side:

Table 3.16: Practical observations.

S. no.	Voltage of primary side, V_1 (V)	Current of primary side, I_o (A)	Power on primary side, P_0 (W)
1	230	4.6	420

On the secondary side:

Table 3.17: Practical observation.

S. no.	Voltage of secondary side, V_{SC} (V)	Current of secondary side, I_{SC} (A)	Power on secondary side, W_{SC} (W)
1	60	5	50

$$\text{Power factor } \cos\phi_p = \frac{W_p}{V_p \times I_p} = \frac{420}{230 \times 4.6} = 0.793$$

$$\text{Power factor } \cos\phi_{sc} = \frac{W_{sc}}{V_{sc} \times I_{sc}} = \frac{50}{60{*}5} = 0.1666$$

$$\text{Iron loss } (W_i) = V_P \times I_p \times \text{Cos}\,\phi_0 = 230 \times (4.6) \times 0.793 = 838.994\ \text{W}$$

$$\text{Iron losses of each transformer } W_I = \frac{W_p}{2} = \frac{420}{2} = 210\ \text{W}$$

$$\text{Copper loss } (W_c) = (V_{SC}) \times I_{SC} \times \cos\phi_{sc} = W_{SC} = 60 \times 5 \times 0.1666 = 49.8\ \text{W}$$

$$\text{Copper losses of each transformer } W_{SC} = \frac{Wsc}{2} = \frac{50}{2} = 25\ \text{W}$$

Total resistance, reactance and impedance are referred to primary:

$$R_{01} = \frac{W_{sc}}{I_{sc}^2} = \frac{50}{5^2} = \frac{50}{25} = 2\,\Omega$$

$$Z_{01} = \frac{V_{sc}}{I_{sc}} = \frac{30}{5} = 6\,\Omega$$

$$X_{01} = \sqrt{(Z_{01})^2 - (R_{01})^2} = \sqrt{(6)^2 - (2)^2} = 4\sqrt{2} = 5.656\,\Omega$$

$$\text{Total losses} = W_I + W_{cu} = 52.5 + 25 = 77.5$$

$$\text{Power} = V \times I$$

(i) Power at full load $= V \times I = 230 \times 4.6 = 1,058$ W

(ii) Power at $\frac{1}{2}$ load $= V \times \frac{1}{2} \times I = 230 \times \frac{1}{2} \times 4.6 = 529$ W

(iii) Power at $\frac{1}{4}$ load $= V \times \frac{1}{4} \times I = 230 \times \frac{1}{4} \times 4.6 = 264.5$ W

(iv) Power at $\frac{3}{4}$ load $= V \times \frac{3}{4} \times I = 230 \times \frac{3}{4} \times 4.6 = 793.5$ W

$$\text{Efficiency, } \%\eta = \frac{\text{output}}{\text{output} + \text{losses}(W_i + W_{cu})} \times 100$$

(i) $\%\eta$ at full load $= \dfrac{1,058}{1,058 + 77.5} \times 100 = 93.17$

(ii) $\%\eta$ at $\frac{1}{2}$ load $= \dfrac{529}{529 + 77.5} \times 100 = 87.22$

(iii) $\%\eta$ at $\frac{1}{4}$ load $= \dfrac{264.5}{264.5 + 77.5} \times 100 = 77.33$

(iv) $\%\eta$ at $\frac{3}{4}$ load $= \dfrac{793.5}{793.5 + 77.5} \times 100 = 91.102$

3.7.3.8 Graph

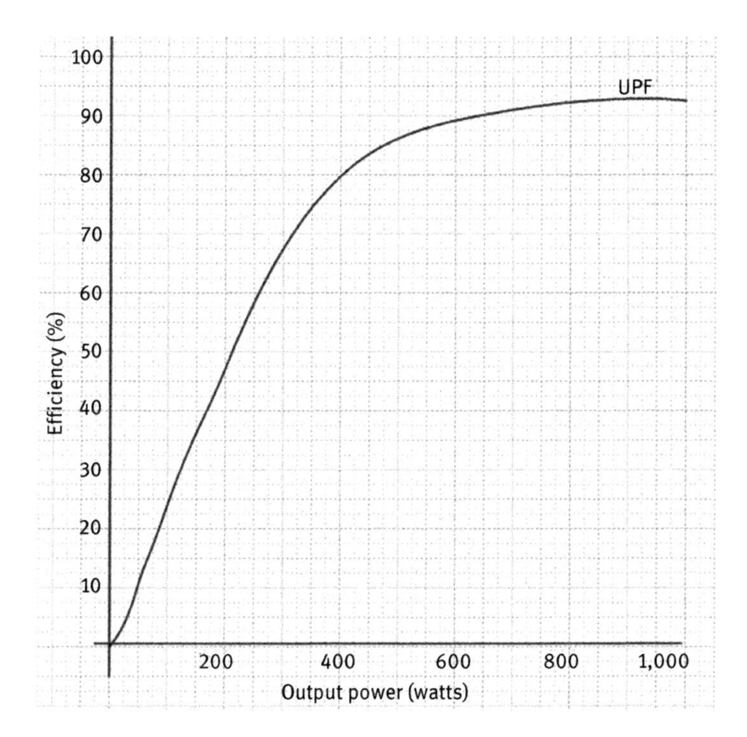

Fig. 3.17: Graph of output power versus efficiency.

3.7.3.9 Result

Performed Sumpner's test and calculated the efficiency of transformers under various power factors and drawn the graph.

3.7.4 Scott connection of transformer

For certain applications, only balanced two-phase supply is required but in general only three-phase supply is available for most of the industries. Hence, there is a need of converting three-phase supply into two phases. It is performed by Scott connection. Using Scott connection of the transformer, three-phase supply is converted into balanced two-phase supply.

3.7.4.1 Apparatus required

Table 3.18: List of apparatus required.

Name	Range	Quantity
Voltmeter	0–400 V, MI	3
Voltmeter	0–200 V, MI	2
Ammeter	0–5 A, MI	5
Two identical T/f	400/230 V, 2 kVA, 50 Hz with 50% and 86.6% tapping	

3.7.4.2 Circuit diagram

Voltmeter method

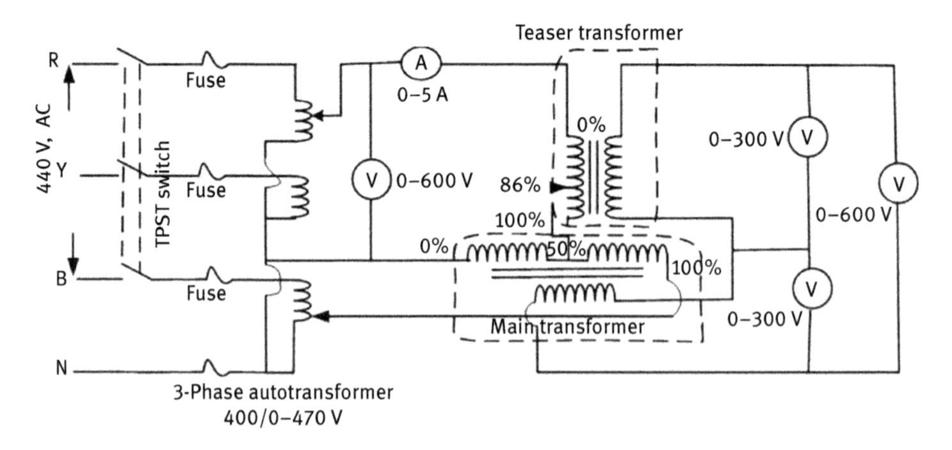

Fig. 3.18: Circuit diagram of Scott connection of transformer: voltmeter method.

3.7.4.3 Theory

The Scott connection of two single-phase transformers is employed for the conversion of a balanced three-phase system into balanced two-phase system or vice versa. Rating of one transformer should be 15% greater than that of the other, but in practice two identical transformers are used for interchangeability and spares. The connection scheme known as Scott connections is shown in Fig. 3.18. In this 50% tap of one transformer (main transformer) is connected to 86.6% tap of the other transformer (teaser transformer). The secondary windings of transformers for balanced supply system have an equal number of turns. The Scott connection of two single-phase transformers with turns ratio of N_1:N_2 is shown in Fig. 3.18. The phase diagram of line voltages on the primary side V_{AB}, V_{BC}, V_{CA} forms an equilateral triangle.

Let

$$V_{AB} = V$$

$$V_{BC} = V\,(-120^0)$$

$$V_{CA} = V(120^0)$$

The secondary voltage of the main transformer is given by

$$V_b = \frac{N_2.V_{CB}}{N_1} = \frac{N_2}{N_1} V\left(60^0\right) \tag{3.10}$$

The voltage V_{AM} is given by

$$V_{AM} = V_{AB} + \frac{V_{BC}}{2}$$

$$V_{AM} = V + \frac{V}{2}(-120^0)$$

$$V_{AM} = \frac{\sqrt{3}}{2} V\,(-30^0)$$

The voltage V_{AM} is across $\frac{\sqrt{3}}{2}$ N_1 turns. Therefore, the primary voltage of teaser transformer is given by

$$V_{AA'} = \frac{2}{\sqrt{3}} V_{AM} = V(-30^0)$$

Hence, the secondary voltage of the teaser transformer is given by

$$V_a = \frac{N_2}{N_1} V_{AA} \tag{3.11}$$

Hence, from eqs. (3.10) and (3.11), for a balanced three-phase supply on the primary side, the voltages on the secondary side of the transformer are equal in magnitude but 90° out of phase. Therefore, a balanced two-phase supply from the balanced three-phase supply using a Scott connection is obtained.

If the secondary load currents are I_a and I_b, then the primary currents can be obtained as follows:

$$I_A = \frac{2}{\sqrt{3}} \frac{N_2}{N_1} I_a$$

$$I_{CB} = \frac{N_2}{N_1} I_b$$

$$I_B = -I_{CB} - \frac{I_A}{2} \text{ and } I_C = I_{CB} - \frac{I_A}{2}$$

To get single-phase voltage supply, short negative polarity side of teaser transformer and positive polarity of the main transformer on the secondary side take the voltage across the positive polarity of teaser transformer and negative polarity of the main transformer on the secondary side. This single-phase voltage is given by

$$V = \sqrt{V_a^2 + V_b^2}$$

3.7.4.4 Procedure

1. The details of the transformer should be noted.
2. The connections of various terminals are connected as per the circuit with Triple Pole Single Throw (TPST) off position.
3. Close the TPST and using three-phase autotransformer, gradually apply the rated voltage in steps to the main transformer.
4. Note down the readings in meters in each step.
5. After completion of experiment bring the autotransformer into zero position.

3.7.4.5 Observation table

Voltmeter method:

Table 3.19: Observation table.

S. no.	Voltmeter reading V_1	Ammeter reading I_1	Voltmeter reading V_{2M}	Voltmeter reading V_{2T}	Voltmeter reading V_{2TM}	$V_{2TM} = \sqrt{V_{2m}^2 + V_{2Tm}^2}$
1						
2						

3.7.4.6 Practical calculations

The Scott connection is verified with 2 kVA Scott-connected transformer and the readings were tabulated.

Tabular form

$$\mathbf{V_{YB} + V_{BR} + V_{RY} = V_1 + V_2}$$

S. no.	V_{RY}	V_{YB}	V_{BR}	V_1	V_2	$V=\sqrt{V_1^2+V_2^2}$
1	400 V	400 V	400 V	230 V	230 V	325.26 V

3.7.4.7 Result

Converted three-phase supply into two-phase supply by using Scott connection and verified.

3.8 Problems

1. A 20 kVA 2,200/220 V 50 Hz distribution transformer is tested for efficiency and regulation and the recorded values are:
   OC test: 220 V, 4.2 A, 148 W–LV side : SC test: 86 V, 10.5 A, 360 W–HV side.
   Determine the efficiency and regulation at 0.8 pf lag at half and full loads.
   Core loss found from the OC test from data is given as 148 W.

From SC test

Full-load current $(I_2) = kVA/V_{HV} = 20,000/2,200 = 9.09A$

$$R_{01} = \frac{P_s}{I_s^2} = \frac{360}{10.5^2} = 3.26\ \Omega$$

Referred to secondary is $R_{02} = K^2R_{01} = \left(\frac{220}{2,200}\right)^2 \times 3.26 = 0.0326\Omega$

$$Z_{01} = \frac{V_{SC}}{I_{SC}} = \frac{86}{10.5} = 8.19\ \Omega$$

$$X_{01} = \sqrt{(Z_{01})^2 - (X_{01})^2} = \sqrt{(8.19)^2 - (3.26)^2} = 7.51\ \Omega$$

Referred to secondary is $X_{02} = K^2X_{01} = \left(\frac{220}{2,200}\right)^2 \times 7.51 = 0.0751\ \Omega$

Efficiency at full load 0.8 pf lag

$$\%\,\text{Efficiency},\ \eta = \frac{\text{kVA} \times \cos\phi}{\text{kVA}\cos\phi + \text{losses}} = \frac{(20,000) \times 0.8}{(20,000) \times 0.8 + 148 + 360} = \frac{16,000}{16,508}$$

$$= 96.92\%$$

$$\text{At } 1/2 \text{ of full load} = \frac{\frac{1}{2}(20,000) \times 0.8}{\frac{1}{2}(20,000) \times 0.8 + 148 + \left(\frac{1}{2}\right)^2 \times 360} = \frac{8,000}{8,000 + 148 + 90}$$

$$= \frac{8,000}{8,238} = 97.11$$

$$\text{Percentage regulation} = \frac{I_2R_{02}\cos\phi \pm I_2X_{02}\sin\phi}{V_2}$$

$$\text{Regulation at full load 0.8 pf lag} = \frac{I_2R_{02}\cos\phi \pm I_2X_{02}\sin\phi}{V_2}$$

$$\frac{9.09(0.0326 \times 0.8 + 0.0756 \times 0.6)}{2,200} = \frac{0.237 + 0.412}{2,200} = \frac{0.649}{2,200} \times 100 = 0.02$$

$$\text{Regulation at 1/2 load 0.8 pf lag} = \frac{9.09/2(0.0326 \times 0.8 + 0.0751 \times 0.6)}{2,200}$$

$$= \frac{0.118 + 0.206}{2,200} = \frac{0.324}{2,200} \times 100 = 0.014$$

2. The separation of losses test is conducted on single-phase transformer and the watt meter reads 63 W at 40 Hz and 110 W at 60 Hz. The test is conducted by maintaining $\frac{V}{f}$ ratio constant. Calculate hysteresis loss and eddy current losses at a frequency of 50 Hz.

Solution:

Core loss at 40 Hz = 63 W.
Core loss at 60 Hz = 110 W.
$\frac{V}{f}$ ratio maintained is constant and the values A and B are calculated using the below relations:

$$\text{Hysteresis loss } P_h = Af$$

$$\text{Eddy current loss } P_e = B f^2$$

$$\text{The total core loss } P_i = P_h + P_e = Af + B f^2$$

$$63 = A \times 40 + B \times 40^2 \qquad \text{...........(1)}$$

$$110 = A \times 60 + B60^2 \qquad \text{...........(2)}$$

Solving eqs. (1) and (2) we get

$$A = 1.0584,\ B = 0.0129$$

At 50 Hz

$$\text{Hysteresis loss } P_h = Af = 1.0584 \times 50 = 52.92\text{W}$$

$$\text{Eddy current loss } P_e = Bf^2 = 0.0129 \times 50^2 = 32.25\text{W}$$

3.9 Viva-Voce questions

OC and SC tests

1. State Faraday's laws.
2. Explain the principle of working transformer.
3. What do you understand the term "core losses"?
4. Why OC and SC are performed?
5. Can the transformer work on DC supply?
6. Explain the procedure to conduct OC test.
7. What is meant by equivalent circuit of the transformer referred to primary?
8. Under no-load condition, draw the phasor diagram of a transformer.
9. How do you minimize the hysteresis losses?
10. Under what condition the regulation of a transformer becomes zero?
11. Define voltage regulation of a transformer.
12. Write the formula for the % efficiency of a transformer.
13. Write the condition for maximum efficiency of a transformer.
14. Define turns ratio.
15. Define transformation ratio of the transformer.

Separation of losses test

16. The core losses depends upon_________.
17. How do you minimize the eddy current losses?
18. Write the emf equation of the transformer.
19. What is autotransformer?
20. Mention the differences between transformer and autotransformer.
21. Why the transformer rating is expressed in kVA?
22. What is meant by tap changing transformer?
23. Mention the applications of transformers.
24. Can 50 Hz transformers be operated at 60 Hz and vice versa?
25. Can transformers be used in parallel?
26. What is meant by regulation in a transformer?
27. What is meant by temperature rise in a transformer?
28. What is meant by "impedance" in transformers ratio?
29. Why V/f ratio is maintained constant?
30. How V/f ratio is maintained constant?

Sumpner's test

31. Why two identical transformers are required for back-to-back test?
32. Mention the advantages of Sumpner's test.
33. Define line and phase voltages and what relation is between them in a star and delta connected system.
34. Why star to delta and delta to star transformation is required?

35. What is the use of breather in a transformer?
36. What is power transformer?
37. Define all day efficiency of a transformer.
38. Why the core is made of silicon steel laminations?
39. What is the role of Buchholz relay?
40. At which load transformer can give maximum efficiency.
41. Differentiate Sumpner's test and OC and SC tests on transformer.

Scott connection

42. What is Scott connection?
43. What is open delta connection?
44. During SC test, if secondary winding is suddenly open-circuited, what will happen?
45. Draw the phasor diagram for Scott connection.
46. Expand C.R.G.O.S.
47. How 6-phase supply is produced? Where it is used?
48. How tap changing transformer works?
49. For which winding the tap changing is provided?
50. Write the applications of Scott connection transformers.

3.10 Objective questions

1. In a test on transformer at 60 V, 30 Hz it gives Φ_1 as lagging power factor. If the same test is performed at 60 V, 50 Hz, then
   (a) $I_1 > I_2$ (b) $\Phi_1 < \Phi_2$ (c) $R_1 = R_2$ (d) $X_1 = X_2$

2. A 2,000/100 V stepdown transformer has $R_1 = 1.5\ \Omega$, $R_2 = 0.005\ \Omega$, $X_1 = 2.5\ \Omega$ and $X_2 = 0.08\ \Omega$. The value of the total resistance referred to primary is
   (a) 6 Ω (b) 6.25 Ω (c) 3.5 Ω (d) 4.82 Ω

3. ________ winding has more number of turns in a transformer
   (a) LV (b) HV (c) Primary (d) Secondary

4. The core losses are negligible in SC test because the applied voltage to the secondary winding is
   (a) zero (b) constant (c) low (d) large

5. The ratings of a transformer are expressed in terms of
   (a) V (b) A (c) kW (d) kVA

6. The hysteresis loss is proportional to
   (a) I_{max} (b) ${B_{max}}^{1.6}$ (c) ${B_{max}}^{3.75}$ (d) $B_{max}/2$

7. A transformer has hysteresis loss of 30 W at 240 V, 60 Hz, the hysteresis loss at 200 V, 50 Hz will be
   (a) 28 W (b) 30 W (c) 25 W (d) 36 W

8. Sumpner's experiment is performed to find out the ________ of the transformer.
   (a) temperature rise
   (b) rating
   (c) dielectric strength
   (d) turns ratio

9. The transformer efficiency is maximum when copper loss is ________ constant losses.
   (a) ≠ (b) > (c) < (d) =

10. Efficiency of a power transformer is __ approximately.
    (a) 90% (b) 98% (c) 95% (d) 92.5%

11. A 230 V, 50 Hz transformer takes a current of 4 A at 0.2 pf lag, then the magnitude of exciting current (I_o) and constant loss (P_c) is
    (a) 3.91 A, 184.1 W
    (b) 3.62 A, 482.4 W
    (c) 9.319 A, 384 W
    (d) 3.33 A, 682 W

12. In a transformer I/P volt amperes =
    (a) O/P volt amperes (b) losses (c) power factor (d) turns ratio

13. Δ/Y-transformer operates suitably when the load is – only.
    (a) balanced (b) unbalanced (c) both a and b (d) none

14. The load sharing between two transformers that are operated in parallel depends on
    (a) reactance
    (b) pu impedance
    (c) losses
    (d) size of the transformer

15. The purpose of the Scott connection of the transformer is to convert__
    (a) 1-Θ to 3-Θ (b) 1-Θ to 2-Θ (c) 3-Θ to 2-Θ (d) all of these

16. The voltage across teaser transformer leads the main transformer supply voltage by
    (a) 120° (b) 180° (c) 90° (d) 10°

17. __ transformer capacity is reduced in open delta-connected transformer in comparison with the normal transformer.
    (a) 26.5% (b) 57.7% (c) 78.9% (d) 67.9%

18. The relation between line (V_L) and phase voltage (Vph) in Δ-connected
    (a) V_L = 1.2 Vph
    (b) V_L = 1.732 Vph
    (c) V_L = Vph
    (d) V_L = 2 Vph

19. The % of trappings between main and teaser transformers
    (a) 50%, 86.6%
    (b) 25%, 25%
    (c) 25.6%, 86.6%
    (d) 66.6%, 66.6%

20. The phase angle difference between three windings in a three-phase transformer is
    (a) 45° (b) 180° (c) 120° (d) 90°

Answers

1. b, 2. c, 3. b, 4. c, 5. d, 6. b, 7. c, 8. a, 9. d, 10. b, 11. a, 12. a, 13. c, 14. b, 15. c, 16. c, 17. b, 18. c, 19. a, 20. C

3.11 Exercise problems

1. The OC and SC test data are given below for a single phase, 5 kVA, 200 V/400 V, 50 Hz transformer.

OC test from LV side:	200 V	1.25 A	150 W
SC test from HV side:	20 V	12.5 A	175 W

   Draw the equivalent circuit of the transformer (i) referred to LV side and (ii) referred to HV side inserting all the parameter values.
2. The parameters of a 2,300/230 V, 50 Hz transformer are given below: R_1 = 0.286 Ω, R_2 = 0.319 Ω, R_O = 250 Ω, X_1 = 0.73 Ω, X'_2 = 0.73 Ω, X_O = 1250 Ω. The secondary load impedance Z_L = 0.387 + j0.29. Solve the exact equivalent circuit across the primary.
3. A 50 kVA, 2400: 240 V transformer has the following test data:

	V	I (A)		P (W)
OC test	240	5.41	Iron loss	186(LV side)
SC test	48	20.8	Copper loss	617(HV side)

   Determine the percentage efficiency and voltage regulation of the transformer at full load, 0.8 pf lagging.

4 Induction Motors

4.1 Introduction

Induction motors are utilized in numerous applications to exchange energy in the country.

In most of the industries induction motors are used for many applications due to its robust construction and low cost, its speed can be controlled easily and has better performance when compared to the other AC motors. The major applications are pumps, conveyors, winders etc.

4.2 Construction and working principle of three-phase induction motor

4.2.1 Construction

Stator

The construction of induction motor is shown in Fig. 4.1. The stator is a motionless part of induction motor, it consists of a number of stampings that are made in the form of slots to provide the stator winding or coils and is connected to three-phase supply. The stator winding is connected either in star (Y) or delta (Δ). The number of coils is selected based on the poles requirement and depends on the speed. The stator slots are attached to the frame or yoke which is made up of cast iron and stator slots are made up of high permeability silicon steel to reduce the hysteresis loss.

If N_s = Synchronous speed

f = Frequency of the supply voltage

P = Number of poles

The synchronous speed, $N_s = \frac{120\,f}{P}$

Rotor

The moving part of the induction motor is called rotor. It consists of laminated core with squirrel cage or wound rotor structure and aluminum or copper conductor or bars fitted in rotor slots. The end connections of rotor bars are short circuited by end rings in squirrel cage rotor and connected to slip rings in wound rotor induction motor. The rotor bars are slightly skewed with rotor slots to reduce magnetic locking and humming noise. The rotor is mounted on shaft and the shaft is supported by ball bearings.

https://doi.org/10.1515/9783110682274-004

Fig. 4.1: Construction of induction motor.

4.2.2 Principle of operation

If a current carrying conductor is placed in magnetic field, a force is exerted on the conductors according the Lorenz's law. This force is proportional to current flowing through conductor, magnetic field strength and length of conductor, that is, $F = Bil \sin\theta$.

If three-phase supply is connected to stator of induction motor, it produces revolving magnetic field which sweeps the rotor conductor. These rotor conductors are cut by the revolving magnetic flux, an emf is induced and current starts flowing through the armature conductors due to the short-circuited armature conductors by the end rings. Hence, these conductors experience force and this force rotates the armature conductors and hence the rotor shaft.

4.3 Advantages and disadvantages

(i) The induction motor has robust construction.
(ii) The cost is very less and requires less maintenance.
(iii) The efficiency of the motor is high.
(iv) The induction motors is inherently self starting.

4.4 Analysis of three-phasesquirrel cage induction motor

4.4.1 Speed of rotating magnetic field

The revolving magnetic field revolves with synchronous speed and it depends upon the frequency of the supply voltage and number of poles.

$$\text{Synchronous Speed, } N_S = \frac{120f}{p}$$

Slip

The difference between the synchronous speed (Ns) and the actual rotor speed (Nr) is called slip. The slip of induction motor is expressed as a percentage of synchronous speed.

$$\text{Percentage of Slip, } (\%S) = \frac{N_S - N_r}{N_S}$$

4.4.2 Rotor frequency under running condition

Under running condition, the frequency of the voltage or current induced in the rotor is

f_r = Slip times of (S) * f
f_r: Rotor frequency
f: Supply frequency

4.4.3 Rotor current and power factor

Equivalent circuit of induction motor
In the equivalent circuit

R_1: Resistance of the stator winding
R_2: Resistance of rotor winding
X_1: Stator reactance
X_2: Rotor reactance
X_m: Magnetizing reactance
R_c: Core loss component

In the rotor conductors, the induced voltage and its frequency is affected by the slip.

The equivalent circuit of the induction motor is shown in Fig. 4.2.

The current drawn by the rotor is given by:

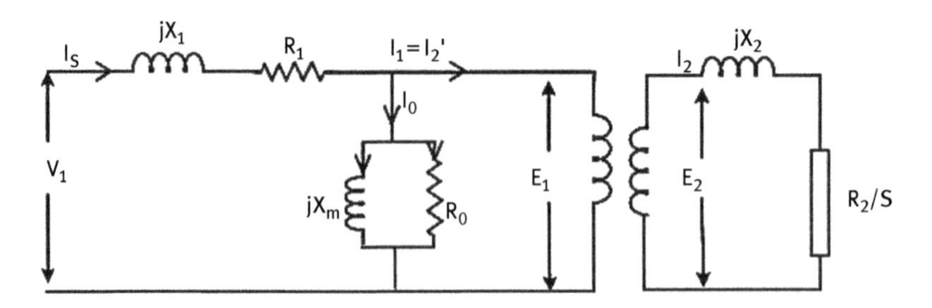

Fig. 4.2: Equivalent circuit of induction motor.

$$I_2 = \frac{SE_2}{\sqrt{R_2^2 + X_2^2}} \tag{4.1}$$

$$I_2 = \frac{E_2}{\sqrt{\left(\frac{R_2}{S}\right)^2 + X_2^2}}$$

$$\text{Power factor} = \frac{R_2}{\sqrt{R_2^2 + (SX_2)^2}} \tag{4.2}$$

Motor input current

The motor input current is calculated by minimizing the equivalent circuit either to stator side or to rotor side. If the equivalent impedance of the motor is Z_{eq}:

$$I_S = \frac{V_1}{Z_{eq}}$$

$$Z_{eq} = R_{eq} + \frac{R_2}{S} + jX_{eq} \tag{4.3}$$

The referred values are calculated by multiplying them by k^2, where "k" is the effective stator/rotor turns ratio.

Motor power input

Power (P_{in}) delivered to the motor per phase is

$$P_{in} = I_S^2\left(R_1 + \frac{R_2}{S}\right) \tag{4.4}$$

Neglected core losses

The copper losses in stator and rotor windings are given by:

$$P_{CU} = I_S^2(R_1 + R_2)$$

The mechanical power developed by the motor per phase is given by:

$$P_m = P_{in} - P_{cu}$$

$$= I_S^2\left(\frac{1-S}{S}\right)R_2 \tag{4.5}$$

$$\text{For 3 phases} = 3\,I_S^2\left(\frac{1-S}{S}\right)R_2 \tag{4.6}$$

$$\text{Efficiency} = \frac{\text{Power developed at the Shaft}}{\text{Power Input to Motor}} = \frac{P_m}{P_{in}} \times 100 \tag{4.7}$$

Torque developed by induction motor under running condition

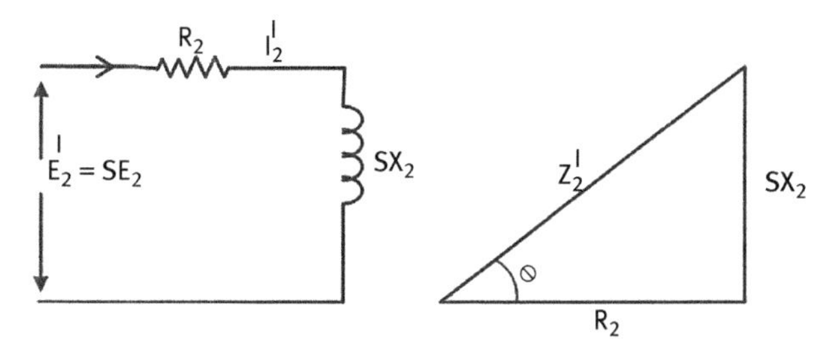

Fig. 4.3: Induction motor equivalent circuit (rotor side).

The torque equation is $T = \frac{3}{\omega}\, E_2^1\, I_2^1 \cos\phi_2$

Where

I_2^1= Current flowing through rotor conductors

E_2^1= Induced voltage in the rotor emf

$\text{Cos}\phi_2$ = Rotor pf

Rotor voltage, $E_2 = SE_2$,

Rotor reactance, $X_2 = SX_2$

$$\text{Rotor current}, I_2 = \frac{SE_2}{Z_2} = \frac{SE_2}{\sqrt{R_2^{\,2} + SX_2^{\,2}}} \tag{4.8}$$

$$\text{Power factor} = \frac{R_2}{\sqrt{R_2^{\,2} + SX_2^{\,2}}}$$

Torque developed by the motor,

$$T = \frac{3}{\omega} E_2\, I_2 \cos\phi_2 = \frac{3}{\omega} E_2 \frac{SE_2}{\sqrt{R_2^{\,2} + SX_2^{\,2}}} \frac{R_2}{\sqrt{R_2^{\,2} + SX_2^{\,2}}} \tag{4.9}$$

It is observed that the torque under running condition is proportional to the square of supply voltage. The equivalent circuit of the inductor motor at the rotor side is shown in Fig. 2.3.

4.5 Testing of three-phase squirrel cage induction motor

4.5.1 Testing on three-phase induction motor-brake test

Brake test is conducted on three-phase squirrel cage induction motor to obtain the performance curves.

4.5.1.1 Apparatus required

Table 4.1: List of required apparatus.

Name	Range	Quantity
Voltmeter	0–600 V (MI)	1
Ammeter	0–10 A (MI)	1
Wattmeter	0–10/20 A, 15, 300, 600 V, UPF	2
Tachometer		1
Connecting wires		

4.5.1.2 Circuit diagram

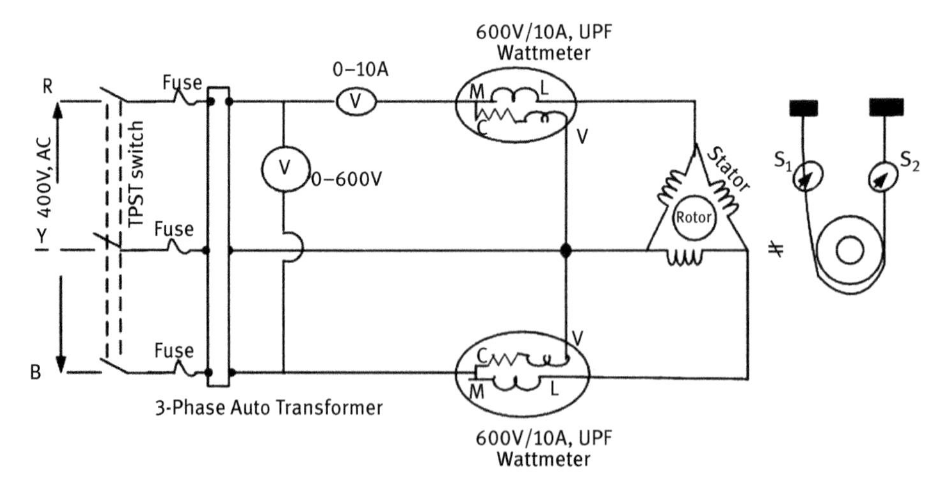

Fig. 4.4: Circuit diagram of three-phase induction motor: brake test.

4.5.1.3 Theory

Induction motor consists of stator and rotor, stator is the stationery part and rotor is the rotating part. The induction motor is also considered as a rotating transformer where transformer secondary is stationary and in induction motor the secondary winding or rotor is rotating. Both induction motor and transformer works on the same principle, that is, electromagnetic induction. Brake test is a direct technique of testing the squirrel cage induction motor in which output power, torque, efficiency and speed are determined at different load conditions.

4.5.1.4 Procedure

1. The name plate details of the induction motor should be noted.
2. The connections of various terminals are connected as per the circuit with TPST off position.
3. Start the induction motor using auto transformer. Apply 30–50 V initially and after attaining rated speed of the motor, apply the rated voltage. (Induction motors draws the high currents during starting.)
4. By increasing the load gradually up to the rated current record the all meter readings.
5. Remove the load gradually and decrease the voltage to zero then switch off the mains.
6. Calculate the necessary parameters to plot the performance curves.

4.5.1.5 Model calculations

Input power drawn by the motor $W(P_i) = (W_1 \pm W_2)$ watts

Shaft Torque $(T) = 9.81 \times (F_2 \sim F_1) \times R$; R is Radius of the drum in mts.

$$\text{Output Power } (P_O) = \frac{2\pi NT}{60}$$

$$\text{Efficiency}(\eta) = \frac{\text{Output Power}(P_O)}{\text{Input Power}(P_i)}$$

$$\%\text{Slip}(s) = \frac{N_s - N_r}{N_s}, \text{Synchronous speed, } N_s = 120\, f/P$$

4.5.1.6 Observation table

Table 4.2: Observation table.

S.No	Voltage (V)	Speed, N (rpm)	Current (A)	Power, W (W) W_1 W_2	Spring balance (kg) F_1 F_2	Torque (Nm)	Output power (W)	% Slip	%η
1									
2									
3									
4									

4.5.1.7 Model graph

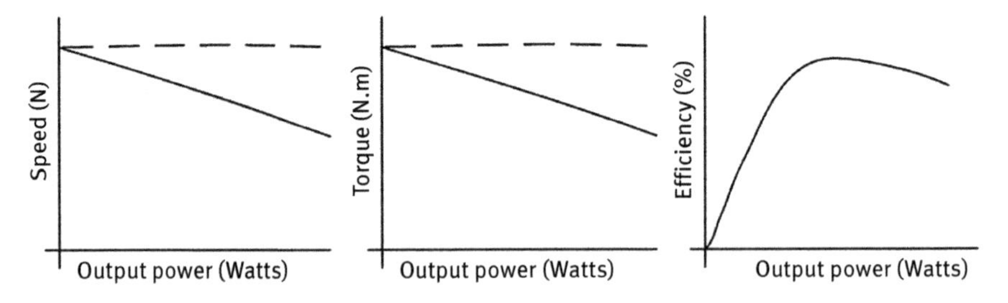

Fig. 4.5: Model graph of three-phase induction motor characteristics.

4.5.1.8 Practical calculations

The brake test is conducted on the following rating of the motor and readings were tabulated.

Rating of the motor = 3.7 kW, 415 V

Current = 7.5 A

Speed = 1500 rpm

Tabular form

Table 4.3: Practical observations.

S.No	Voltage (V)	Speed, N rpm	I (A)	Power, W (W) W_1	W_2	Spring balance (kg) F_1	F_2	Torque (Nm)	Output power (W)	% Slip	%η
1	320	1500	6	20	160	0	0	0	0	0	0
2	320	1495	6.5	65	680	1	5	4	626.22	0.33	84.05
3	320	1492	7	70	800	1.5	7	5.5	859.33	0.53	98.77
4	320	1490	7.25	100	920	2	8.5	6.5	1014.21	0.66	99.43

Input power (W) = $W_1 + W_2$

(i) $W = 20 + 160 = 180$
(ii) $W = 65 + 68 = 745$
(iii) $W = 70 + 800 = 870$
(iv) $W = 100 + 920 = 1210$

Torque (T) = $9.81 \times r \times (F_2 - F_1)$

(i) $T = 9.81 \times 0.102 \times (0) = 0$
(ii) $T = 9.81 \times 0.102 \times (4) = 4$
(iii) $T = 9.81 \times 0.102 \times (5.5) = 5.5$
(iv) $T = 9.81 \times 0.102 \times (6.5) = 6.5$

Output Power $= \dfrac{2\pi NT}{360}$

i) $\text{Output Power} = \dfrac{2 \times \pi \times 1500 \times 0}{60} = 0$

ii) $\text{Output Power} = \dfrac{2 \times \pi \times 1495 \times 4}{60} = 626.22\,\text{W}$

iii) $\text{Output Power} = \dfrac{2 \times \pi \times 1492 \times 5.5}{60} = 839.33\,\text{W}$

iv) $\text{Output Power} = \dfrac{2 \times \pi \times 1490 \times 6.5}{60} = 1014.21\,\text{W}$

Efficiency, $\%\eta = \dfrac{\text{Output}}{\text{Input}} \times 100$

i) $\%\,\text{Efficiency}, \eta = \dfrac{0}{180} \times 100 = 0$

ii) $\% \text{Efficiency}, \ \eta = \frac{626.22}{745} \times 100 = 84.05$

iii) $\% \text{Efficiency}, \eta = \frac{859.33}{870} \times 100 = 98.77$

iv) $\% \text{Efficiency}, \eta = \frac{1014.21}{1020} \times 100 = 99.43$

$$\%\textbf{Slip} = \frac{\text{Ns} - \text{Nr}}{\text{Ns}} \times 100$$

i) $\frac{1500 - 1500}{1500} \times 100 = 0$

ii) $\frac{1500 - 1495}{1500} \times 100 = 0.33$

iii) $\frac{1500 - 1492}{1500} \times 100 = 0.53$

iv) $\frac{1500 - 1490}{1500} \times 100 = 0.66$

The performance curves, that is, speed, torque and efficiency versus output power are drawn.

4.5.1.9 Result

Brake test is conducted on three-phase induction motor and drawn the performance curves, that is, speed, torque and efficiency versus output power.

4.5.2 No-Load and blocked rotor test on three-phase induction motor

The performance parameters of the three-phase induction motor is evaluated by using circle diagram.

4.5.2.1 Apparatus required

Table 4.4: List of required apparatus.

Name	Range	Quantity
Voltmeter	0–600 V, MI	1
Ammeter	0–10 A, MI	1
Three-phase variac	415/0–470 V	1
Wattmeter	0–600 V, 0–10 A, LPF	2
Connecting wires & tachometer		

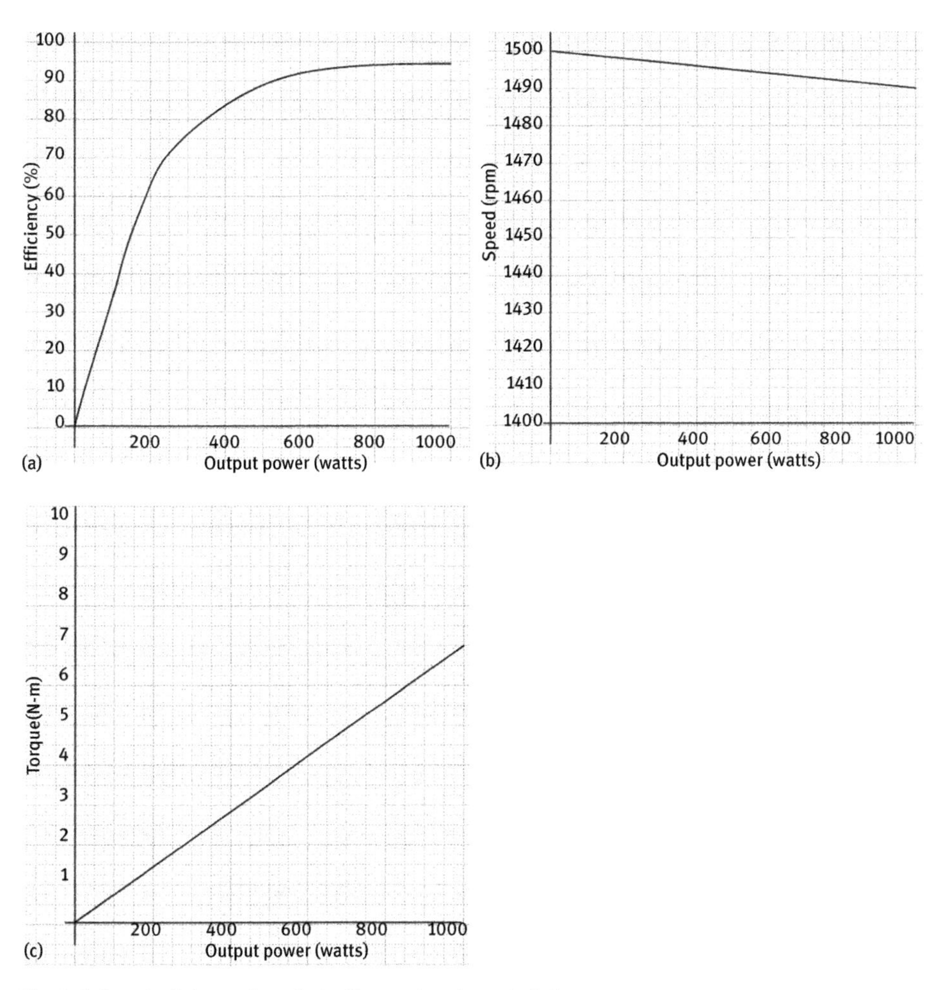

Fig. 4.6: Graph of three-phase induction motor characteristics.

4.5.2.2 Circuit diagram

4.5.2.3 Theory

No-load and blocked rotor test is performed on induction motor to draw the circle diagram from which power factor, losses and input current are measured without any mathematical calculations. Therefore, circle diagram is very useful to analyze the performance of the induction motor. To draw the circle diagram, the induction motors are operated under no-load and short-circuit conditions.

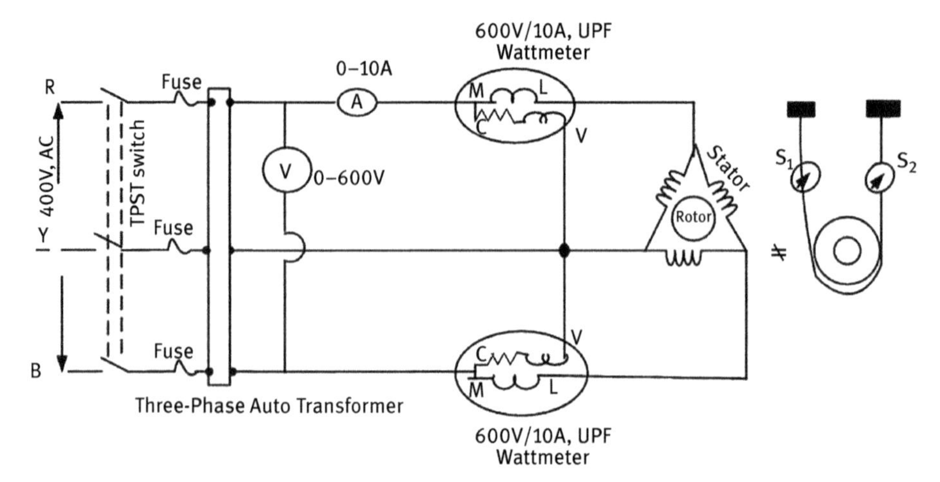

Fig. 4.7: Circuit diagram of three-phase induction motor: no-load and blocked rotor test.

4.5.2.4 Procedure

No-load test

1. The name plate details of the induction motor should be noted.
2. The connections of various terminals are connected as per the circuit with TPST off position.
3. Start the induction motor using auto transformer. Apply 30–50 V initially and after attaining rated speed of the motor, apply the rated voltage.
4. Without applying any load, record all meter readings and bring the three-phase auto transformer to "0" volts position and switch of the motor.

Blocked rotor test

5. Now tighten the motor pulley using belt till blocked rotor condition is reached.
6. Start the induction motor using auto transformer. Slowly increase supply voltage till the motor draws the rated short-circuit current, that is, blocked rotor current.
7. Remove the load gradually and decrease the voltage to zero, then switch off the mains.
8. Draw the circle diagram and measure the required parameters.

4.5.2.5 Observation table

No-load test

Table 4.5: Observation table.

S.No.	Voltmeter reading, V_o (V)	Ammeter reading, I_o (A)	Wattmeter reading		W_o (P_o) $W_1 + W_2$(W)
			W_1 (W)	W_2 (W)	
1					

Blocked rotor test

Table 4.6: Observation table.

S.No.	Voltmeter reading, V_{sc} (V)	Ammeter reading, I_{sc} (A)	Wattmeter reading		W_{sc} (P_{sc}) $W_1 + W_2$ (W)
			W_1 (W)	W_2 (W)	
1					

4.5.2.6 Circle diagram

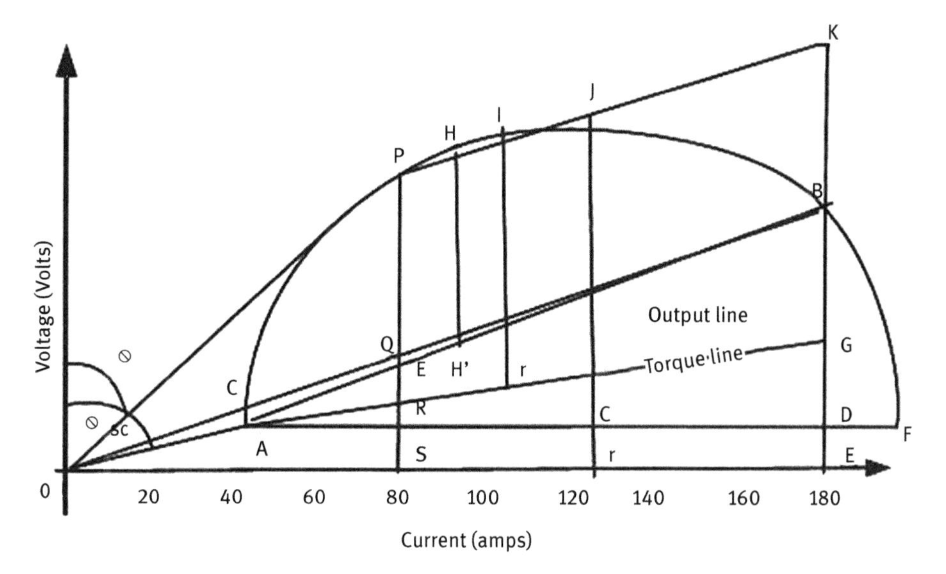

Fig. 4.8: Model graph of circle diagram of three-phase induction motor.

4.5.2.7 Procedure for construction of circle diagram

The circle diagram is constructed with values obtained from no-load and blocked rotor test with the help of below procedure:

1. Consider current phasor on X-axis and voltage phasor on Y-axis as reference vectors.
2. Choose appropriate current range such that the diameter of a circle is 20–30 cm.
3. No-load power factor (Cos ϕ_O) is calculated from the no-load test. Draw the vector I_O, lags V by an angle ϕ_O. It is line OA.
4. From point "A" draw a parallel line to the horizontal axis.
5. Full-load power factor (Cos ϕ_{sc}) is calculated from the blocked rotor test. Draw the vector I_{sc}, lags V by an angle ϕ_{sc} with the same scale. It is line OB.
6. To get the output line, connect the points AB,
7. To locate the center, bisect the line AB and extend it to meet line AD. Point C is called the center of the circle.
8. With center "C" draw a circle with a radius AC. This meets the horizontal line drawn from A at B.
9. Draw the perpendicular from point B on the horizontal axis to meet AF line at D and meet the horizontal axis at E.

Torque line (AG)

The torque line separates stator and rotor copper losses.

The vertical distance BD represents power input at short circuit, that is, W_{SN}, which consists of core loss, stator and rotor copper losses.

FD = DE = fixed loss

AF = Sum of stator and rotor copper losses

Point "G" is located as

$$\frac{BG}{GD} = \frac{\text{Rotor Copper Loss}}{\text{Stator Copper Loss}}$$

Power scale

As AD represents W_{SN}, that is, power input on a short circuit at normal voltage, the power scale can be obtained as

$$\text{Power scale} = \frac{W_{SN}}{L(BE)}$$

L(BE) = Distance BE in cm

Location of point E (slip ring induction motor)

$$K = \frac{I_2}{I_1} = \text{Transformation ratio}$$

$$= \frac{AE}{BF} = \frac{\text{Rotor Cu loss}}{\text{Stator Cu loss}} = \frac{R_2}{R_1}\left(\frac{I_2}{I_1}\right)^2$$

$$R_2{}^1 = \frac{R_2}{K_2} = \text{Rotor resistance referred to stator}$$

$$\frac{BG}{GD} = \frac{R_2{}^1}{R_1}$$

Thus, point G can be obtained by dividing the line BD in the ratio $R_2{}^1$ and R_1.

Location of point D (squirrel cage induction motor)

The stator resistance is calculated by using resistance test.

That is, stator copper loss = $3I^2_{SN}{}^{*}R_1$: where I_{SN} is phase value

Neglecting core loss, W_{SN} = stator Cu loss + Rotor Cu loss

That is, rotor copper loss = $W_{SN} - 3I^2_{SN}R_1$

$$\frac{BG}{GD} = \frac{W_{SN} - 3I_{SN}{}^2R_1}{3I_{SN}{}^2R_1}$$

Dividing line BD in this ratio, the point G can be obtained and hence AG represents torque line.

To get the torque line, join the points A and G.

10. To find the full-load quantities, draw line BK (=Full-load output/power scale). Now, draw line PK parallel to output line meeting the circle at point P.
11. Draw line PT parallel to Y-axis meeting output line at Q, torque line at R, constant loss line at S and X-axis at T.

4.5.2.8 Practical calculations

No-load and blocked rotor test is conducted on the following rating of the induction motor and recorded data.

Rating of the machine: 3.5 kW

Rated Speed: 1500 rpm

Full-load current: 22 A

No-load test

Table 4.7: Practical observations.

S.No.	Voltmeter reading, V_o (V)	Ammeter reading, I_o (A)	Wattmeter reading		W_o (P_o) W_1+W_2 (W)
			W_1 (W)	W_2 (W)	
1	320	6	20	160	180

Blocked rotor test

Table 4.8: Practical observations.

S.No.	Voltmeter reading, V_{sc} (V)	Ammeter reading, I_{sc} (A)	Wattmeter reading		W_{sc} (P_{sc}) $W_1 + W_2$ (W)
			W_1 (W)	W_2 (W)	
1	90	7.5	80	280	360

At no load

Output Power, $(W_O) = \sqrt{3}V_O I_O\cos\phi_O = 180 \times 320 \times 6 \times \cos\phi_O$

$$\cos\phi_O = \frac{180}{1.732 \times 320 \times 6} = 0.0541$$

$$\phi_o = \text{Cos}^{-1}(0.0541) = 86.89^0$$

At SC test (on blocked rotor test)

$$W_{SC} = \sqrt{3}\,V_{SC}\,I_{SC}\,\cos\phi_{sc}$$

$$360 = \sqrt{3} \times 90 \times 7.5 \times \cos\phi_{sc}$$

$$\cos\phi_{sc} = 0.3079$$

$$\phi_{sc} = \text{Cos}^{-1}(0.3079) = 72.06^0$$

$$I_{SN} = I_{SC} \times \left[\frac{V_l}{V_{SC}}\right] = 7.5 \times \left[\frac{320}{90}\right]$$

$$I_{SN} = 26.66\text{A}$$

$$W_{SN} = W_{SC} \times \left[\frac{I_{SN}}{I_{SC}}\right]^2 = 360 \times \left[\frac{26.66}{7.5}\right]^2 = 4548.33\text{W}$$

$$\%\,\text{Efficiency} = \frac{\text{Output}}{\text{Input}} \times 100$$

From circle diagram

From no-load test

$$\cos\phi_O = 0.09, \phi_O = 84.62^\circ$$

From SC or blocked rotor test

$$\cos\phi_{sc} = 0.307,\ \phi_{sc} = 72.06^\circ, I_{SN} = 26.66\text{A}, W_{SN} = 4548.33\text{W}$$

$$\text{Power Scale} = \frac{\text{Wsn}}{\text{L(AD)}} = \frac{4548.33}{2.6} = 1749.35$$

$$AA' = \frac{3 \times 103}{1749.35} = 1.714 \text{W/cm}$$

4.5.2.9 Result

Circle diagram of three-phase induction motor is drawn by conducting no-load and blocked rotor test.

4.5.3 Testing of single-phase induction motor-equivalent circuit parameters

The no-load and blocked rotor test is conducted on a single-phase induction motor, the equivalent circuit parameters are evaluated.

4.5.3.1 Apparatus required

Table 4.9: List of required apparatus.

Name	Range	Quantity
Voltmeter	0–300 V(MI)	1
Ammeter	0–2/10 A(MI)	1
Wattmeter	0–5/10 A, 300 V, LPF	2
Three-phase auto T/f	(415/0–470 V) dimmer	1

4.5.3.2 Circuit diagram

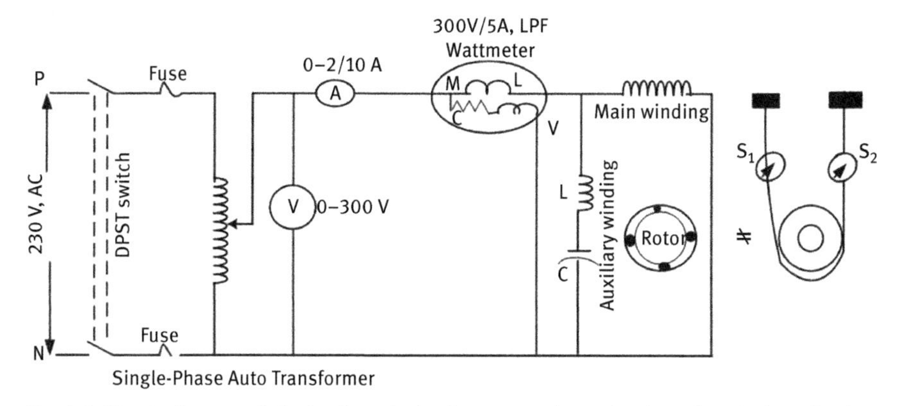

Fig. 4.9: Circuit diagram of single-phase induction motor: determination of equivalent circuit parameters.

4.5.3.3 Theory

The constructional features of single-phase induction motors are similar to three-phase induction motor but it is not self-starting because revolving magnetic field is not produced. Therefore, single-phase induction motors are constructed with two windings, namely starting winding and running winding, which produce revolving magnetic field due to the resultant flux produced by the starting winding flux and running winding flux. Once the motor attains its rated speed, the starting winding is disconnected to reduce the losses, hence efficiency is improved. These motors are also called split phase motors.

The equivalent circuit parameters are shown in Fig. 4.10 by neglecting iron losses. The stator winding has resistance R_1 and leakage reactance X_1 and the rotor has resistance R_2 and leakage reactance X_2. If the rotor rotates with a slip "S" in forward direction and it has a slip "2-S" in backward direction.

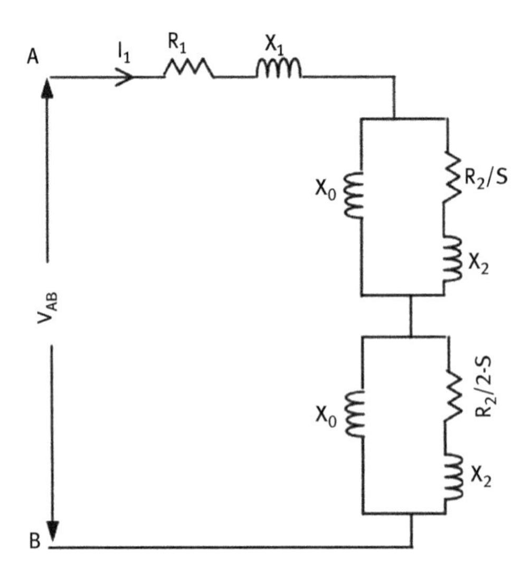

Fig. 4.10: Equivalent circuit of single-phase induction motor.

4.5.3.4 Procedure

DC resistance measurement

The DC resistance R_L of the stator winding is measured using the ammeter voltmeter method.

R_1 = voltmeter reading/ammeter reading

No-load test

1. The name plate details of the induction motor should be noted.
2. The connections of various terminals are connected as per the circuit with DPST off position.
3. Start the induction motor using auto transformer. Apply 30–50 V initially and after attaining rated speed of the motor, apply the rated voltage.
4. Without applying any load, record the all meter readings and bring the single-phase auto transformer to "0" volts position and switch off the motor.

Blocked rotor test

5. Now tighten the motor pulley using belt till blocked rotor condition is reached.
6. Start the induction motor using auto transformer. Slowly increase supply voltage till the motor draws the rated short circuit current, that is, blocked rotor current.
7. Remove the load gradually and decrease the voltage to zero then switch off the mains.
8. Calculate the necessary parameters and draw the equivalent circuit.

Note: Rotor also can be blocked by just opening the starting winding.

4.5.3.5 Observation table

No-load test

Table 4.10: Observation table.

S. No	No-load applied voltage, V_0 (V)	No-load current, I_0 (A)	No-load power, W_0 (W)	Speed (rpm)
1				

Blocked rotor test

Table 4.11: Observation table.

S.No	Blocked rotor voltage, V_{sc} (V)	Blocked rotor current, I_{sc} (A)	Power,W_{sc} (W)
1			

4.5.3.6 Model calculations to draw the equivalent circuit

Blocked rotor test	No-load test
$Z_{eq} = V_{sc}/I_{sc}$	$W_0 = V_0 I_0 \cos\phi_0$
$R_{eq} = W_{sc}/(I_{sc})2$	$\cos\phi_0 = W_0/V_0 \times I_0$
$X_{eq} = \sqrt{(Z_{eq}^2 - R_{eq}^2)}$	$Z_0 = (r_1 + R_2/2) + j(X_1 + X_2)$
$r_1 = 1.5 \times R_{dc}$	$V_{AB} = V_o - I_o \times Z_o$
$R_{eq} = r_1 + R_2$	$V_{AB} = I_0 \times X_0$
$R_2 = R_{eq} - r_1$	$X_0 = V_{AB}/I_0$
$X_{eq} = X_1 + X_2;\ X_1 = X_2$	$\%\ \text{Slip}\ (S) = \frac{Ns - Nr}{Ns} \times 100$
$x_2 = X_2/2;\ r_2 = R_2/2$	

Where

V_{SC} = Short circuit voltage in volts	W_0 = No-load input power in watts
I_{SC} = Short circuit current in amps	V_0 = Line voltage on no-load in volts
W_{SC} = Short circuit power in watts	I_0 = Line current on no-load in amps

4.5.3.7 Equivalent circuit of single-phase induction motor

4.5.3.8 Practical calculations

This test is conducted on a single-phase induction motor with below ratings and readings recorded:

Rating of the Motor: 3.7 kW, 240 V, 1500 rpm, current = 10.5 A

Tabular form

No-load test

Table 4.12: Practical observations.

S.No	No-load applied voltage, V_0 (V)	No-load current, I_0 (A)	No-load power, W_0 (W)	Speed (rpm)
1	230	5.9	280	1450

Blocked rotor test

Table 4.13: Practical observations.

S.NO	Blocked rotor voltage, V_{sc} (V)	Blocked rotor current, Isc (A)	Power,Wsc (W)
1	75	8	400

No-load power, $W_0 = V_0 I_0 \text{Cos}\phi_0$ watts

$$\cos\phi_0 = \frac{W_0}{V_0 I_0} = \frac{280}{230 \times 5.9} = 0.206$$

$$\phi_0 = \cos^{-1}(0.206) = 78.11^\circ$$

$$\sin\phi_0 = \sin(78.11) = 0.97^\circ$$

$$\text{Impedance}\, Z_0 = \frac{V_0}{I_0} = \frac{230}{5.9} = 38.98\Omega$$

Reactance $X_0 = Z_0 \sin\phi_0 = 38.98 \times 0.97 = 37.81\Omega$

From no-load test

$$X_M = 2\left[\frac{V_0}{I_0}\right] = 2 \times 38.98 = 77.96\Omega$$

From blocked rotor test

$$W_{SC} = 400\text{W}, I_{SC} = 8\text{A}, V_{SC} = 75\text{V}$$

$$W_{SC} = V_{SC}\ I_{SC}\ \cos\phi_{sc}$$

$$\cos\phi_{sc} = \frac{W_{sc}}{V_{sc} I_{sc}} = \frac{400}{8 \times 75} = 0.66$$

$$\phi_{sc} = \cos^{-1}(0.66) = 48.7^\circ$$

$$\sin\phi_{sc} = \sin(48.7) = 0.75^\circ$$

$$R_{SC} = Z_{SC} \times \cos\phi_{sc}$$

$$Z_{SC} = \frac{V_{sc}}{I_{sc}} = 9.375\Omega$$

$$9.375 \times 0.66 = 6.187\Omega$$

$$X_{SC} = Z_{SC} \times \sin\phi_{sc} = 9.375 \times 0.75 = 7.03\Omega$$

$$R_{SC} = r_1 + r_2$$

$$R_2 = R_{SC} - r_1$$

$$\text{Where } r_1 = 1.2R_a(R_a = 2.8\Omega)$$

$$= 1.2 \times 2.8 = 3.36\Omega$$

$$R_2 = R_{SC} - r_1$$

$$6.187 - 3.36 = 2.827\Omega$$

$$X_{SC} = X_1 + X_2$$

$$X_1 = X_2 = \frac{X_{sc}}{2} = \frac{7.03}{2} = 3.515$$

$$\text{Slip}(S) = \frac{Ns - N}{Ns} \times 100 = \frac{1500 - 1450}{1500} \times 100 = 3.33\% = 0.03$$

4.5.3.9 Result

The parameters of equivalent circuit of a single-phase induction motor are obtained by conducting no-load and blocked rotor test.

4.6 Problems

1. A 40 HP, 400 V, 4 pole, Δ-connected three-phase induction motor provides the following test data.
   No-load data: 400 V, 20 A, 1200 W
   Blocked rotor data: 100 V, 45 A, 2800 W
   Draw the circle diagram for the aforementioned data.
   Assume Stator, Rotor Cu-losses are equal at standstill.
   Assume all the data given in line-line values.
   Output Power (W_0) = $\sqrt{3}V_0\, I_0\cos\phi_0$

   $$\cos\phi_0 = \frac{1200}{1.732*400*200} = 0.0866$$

   $$\phi_o = 85^o$$

   Similarly, power factor under blocked rotor condition $= \dfrac{2800}{1.732 \times 100 \times 45} = 0.359$
   $\phi_s = 69^o$
   Input current and input power at blocked rotor condition is at a reduced voltage of 100 V.
   These quantities are to be converted into a rated voltage of 400 V.
   Thus, Input current at blocked rotor condition at 400 V.

$$= 45 \times \frac{400}{100} = 180\text{A}$$

Input power at blocked rotor condition at 400 V $= 2800\left(\frac{400}{100}\right)^2 = 44800$ V

No-load power input to the motor at 400 V = 1200 W

Current scale 1 cm = 10 A is considered to construct the circle diagram.

Steps to draw the circle diagram

Current and voltage are represented On X-axis and Y-axis respectively.

Current scale 10 A–1 cm

Then for 20 A–2 cm and no-load angle is 85^0.

The phasor OC represents the no-load current (I_o) lags the voltage by 85^0

The full-load short-circuit current (I_{SC}) 180 A with a length of 18 cm is drawn and it is lagging the voltage phasor by 69^0. The vector CD having a length of 18 cm represent the I_{SC}.

Draw a horizontal line CF from C parallel to the line OE. Join OD. Draw a perpendicular bisector from CD to cut the horizontal line CF at Q. With Q as centre and QC as radius, draw a semicircle. From D on the semicircle, draw a vertical line DM on the horizontal axis. DM cuts CF at N such that NM = CL. Since stator and rotor I^2R-losses are assumed to be equal, divide DN at R and join CR. Now CD represents the output line and CR represents the torque line. In the circle diagram, length DM represents input power at blocked rotor condition, such that DM = 44800 W, DM = 6.8 cm

From this,

$$1\text{cm} = \frac{44800}{6.8} = 6588\text{ W}$$

This is the power scale for the circle diagram.

Now, full load output of the motor = 40 hp

$$= 40 \times 735.5 = 29420\text{ W}$$

On the power scale, 29420 W represent $= \frac{29420}{6588} = 4.46$ cm

The line JP on the top of output line has a length of 4.46 cm. To establish the point of JP, extend vertically the line MD and cut 4.46 cm from it.

The line DG is 4.46 cm in Fig. 4.11. Now sketch a line from G parallel to line CD to meet the circle at P.

From point P draw a line vertically downwards, PJ, onto the point CD.

Then join OP, this line represents the full-load input current.

Input current, OP = 6.1 cm × 10 A = 61 A

Full-load power factor = Cos 34^0 = Cosϕ; $\phi = 0.829^0$

Input power = PU = 5.1 cm × power scale = 5.1 × 6588 W = 33.5 kW

Output power = PJ = 4.46 cm × power scale = 4.46 × 6588 W = 29.38 kW

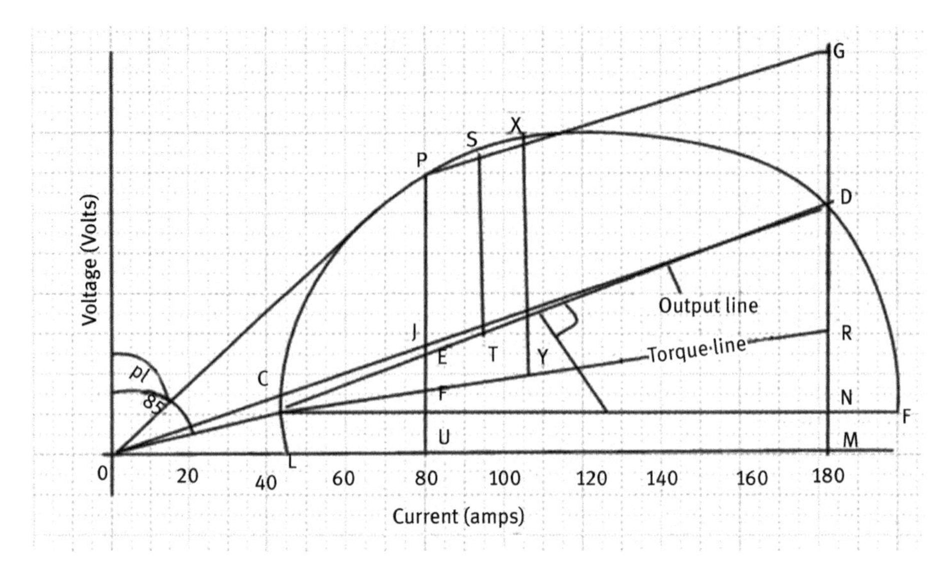

Fig. 4.11: Circle diagram.

$$\textbf{Efficiency} = \frac{\text{Output Power}}{\text{Input Input Power}} = \frac{PJ}{PU} = \frac{4.46}{5.1} = 87.4\%$$

Maximum torque = 6588 × 8.1 Syn. W = 53363 Syn. W.

To determine full-load rotor speed, we can use the relation

Rotor I^2R–loss = Slip × Rotor Input

At full-load rotor I^2R-loss is represented by the distance JK which is equal to 0.3 cm. Rotor input is represented by PK and is equal to 4.65 cm.

Thus, Slip $S = \frac{0.3}{4.65} = 0.0646$

Rotor speed (N_r) = N_S (1-S) = 1500 (1-0.0646) = 1403 rpm

4.7 Viva-voce questions

Brake test on three-phase induction motor

1. The six terminals of three-phase induction motor will be connected in star or delta. How?
2. What is meant by rotating magnetic field?
3. What is the difference between squirrel cage induction motor and slip ring induction motor?
4. Why should the rotor winding of a slip ring induction motor be connected in star?
5. Explain the advantages and disadvantages of slip ring induction motor.

6. What is the use of slip rings in a slip ring induction motor?
7. Is a slip ring induction motor a self-start motor or not?
8. Draw the performance characteristics of a slip ring induction motor.
9. Calculate speed of an induction motor for (a) 3% slip (b) 0% slip?

No-load and blocked rotor test

10. Explain what is meant by a three-phase induction motor.
11. Write the classification of three-phase induction motor.
12. State the steps to draw the equivalent circuit of three-phase induction motor.
13. State the condition for maximum torque of three-phase induction motor.
14. Give the different methods of speed control of induction motor.
15. How do you calculate slip speed?
16. State the condition when the induction motor acts as an induction generator.
17. Give the other name for induction generator
18. Mention the parameters that can be calculated from the no-load test of induction motor?
19. Define slip power of an induction motor.
20. The value of the no-load current in a three-phase induction motor is ——.
21. Under no-load, what is the reading of watt meter?
22. Define the slip of the induction motor.
23. How to obtain calculate power using two-wattmeter method.
24. If single-phase supply is connected to a three-phase induction motor, will the motor start?
25. Define torque line.
26. Draw the torque-slip characteristics of induction motor.
27. How do you calculate real and reactive powers in no-load and blocked rotor test?
28. Discuss the losses in an induction motor.
29. The rotation of three-phase induction motor is to be changed. How?
30. In a star connected three-phase induction motor, the star point is not connected to a neutral of the supply. Why?
31. What is the advantage of a star-delta starter when compared to DOL starter?
32. For P = 4, f = 50 Hz, the synchronous speed is ——.

4.8 Objective questions

1. The "cogging" phenomenon is observed in —— motors.
   (a) synchronous (b) shunt (c) series (d) induction

2. When the rotor of a three-phase induction motor is blocked, its rotor frequency will become
   (a) high
   (b) three forth of input frequency.
   (c) equal to supply frequency
   (d) very low.

3. A three-phase induction motor has 4% slip. The frequency of rotor emf will be
   (a) 5 (b) 2 (c) 4 (d) 1

4. What happens when the induction motor runs at synchronous speed, the motor will
   (a) burn up
   (b) stops
   (c) produce active power
   (d) produce reactive power

5. The no-load value of the slip in an induction motor
   (a) <1% (b) =2.5% (c) =5% (d) =0%

6. A three-phase 440 V, 50 Hz induction motor has 4% slip. The frequency of rotor emfwill be
   (a) 200 Hz (b) 50 Hz (c) 2 Hz (d) 0.2 Hz

7. In brake test, under no-load condition of an induction motor, the readings of two watt meters
   (a) Both wattmeter reads zero
   (b) One will read positive and other reads negative
   (c) Does not measure
   (d) None

8. An induction motor with 1000 rpm speed will have —— poles
   (a) 4 (b) 6 (c) 12 (d) 8

9. —— represents rotor current in an induction motor circle diagram.
   (a) Half the diameter of the circle
   (b) Diameter of the circle
   (c) Two times the diameter of the circle
   (d) Square of the diameter of the circle

10. The circle diagram can be used to find
    (a) losses (b) input power (c) slip (d) all

11. The no-load power factor value in a circle diagram
    (a) <0.5 (b) >0.5 (c) = to 0.5 (d) 1

12. In a single-phase induction motor the net torque produced during the starting is
    (a) high (b) depends of poles (c) low (d) zero
13. —— motors are self starting inherently.
    (a) Single phase (b) Shaded-pole (c) Reluctance (d) All
14. The operating power factor of 1-ϕ induction motor
    (a) 0.6 lagging (b) 0.6 leading (c) 0.8 lagging (d) 0.8 leading
15. The power rating of a fan
    (a) 20 to 25 W (b) 50 to 100 W (c) 50 to 150 W (d) 20–100 W

Answers

1. d, 2. c, 3. b, 4. b, 5. a, 6. c, 7. b, 8. b, 9. b, 10. d, 11. a, 12. d, 13. b, 14. a, 15. c

4.9 Exercise problems

1. A 4-pole, 230 V, three-phase induction motor has a value of secondary resistance such that the motor produces maximum developed torque at stall. Neglect core losses and use the Thevenin's equivalent circuit for analysis. Known equivalent circuit values are:

   $R_1 = 0.2$ $R_2' = 1.1064$

   $X_1 = 0.5\Omega$ $X_m = 20\Omega$

   Find (a) the reflected value of X_2 and (b) the total developed torque at stall.
2. A three-phase, 4–pole, 600 V, 60 Hz induction motor is modeled by $Z_{th} = 0.6933 + j1.933\Omega$ $R_2' = 4.5\Omega$ and $X_2' = 2\Omega$. Find the shaft speed at which maximum torque occurs if the motor is absorbing power from the three-phase lines at rated frequency.
3. A three-phase, 230 V induction motor is operating at no-load condition with rated voltage applied. Equivalent circuit parameters are

   $R_1 = 0.26\Omega$ $R_2' = 0.4\Omega$ $R_C = 143\Omega$

   $X_1 = 0.6\Omega$ $X_2' = 1.4\Omega$ $X_m = 22.2\Omega$
4. It is known that the rated voltage core losses are equal to the rotational losses. Assume that for this no-load condition the coil resistive voltage drops and the leakage reactance voltage drops can be neglected. Determine (a) no-load slip, (b) no-load input power factor and (c) no-load line current.

5 Synchronous Generators

5.1 Introduction

In all types of electric power generating stations, the synchronous generator are installed to generate the AC power. Similar to all electrical machines, it also consists of stator and rotor and works on the principle of faradays laws of electromagnetic induction.

These are also called alternator. These generators run at synchronous speed (120f/p) and are generally 3-phase, since all industrial motors require three-phase supply, and power transmission and distribution is carried out by three-phase system only. Large size alternators are employed in large size hydroelectric and nuclear power stations.

The largest size used in India has a rating of 500 MVA generators employed in superpower thermal power stations and its stator winding is designed for voltage ranging from 6.6 kV to 33 kV.

5.2 Construction and working of synchronous generators

5.2.1 Construction

The main parts of an alternator are namely (i) stator (ii) rotor.

Stator

The stator is made with cast iron for low rating machines and for high rating welded steel is used. Stator core is laminated to minimize core losses and it is made of silicon steel.

A three-phase winding is wounded in the slots as shown in Fig. 5.1. Distributed winding is used for each phase because it produces almost a pure sinusoidal.

Rotor

Alternators employ two types of rotors namely salient pole type and nonsalient pole type as shown in Fig. 5.2.

Salient pole rotor

In salient pole type rotor, the poles are projecting outside from the core of the rotor surface and offer nonuniform air gap. These types of rotors are employing concentric winding and damper bars are in the rotor pole faces to damp out the rotor oscillations during the rapid changes in the load. The rotor pole faces are constructed to produce a sinusoidal voltage. This type of rotors have many poles to operate at low

https://doi.org/10.1515/9783110682274-005

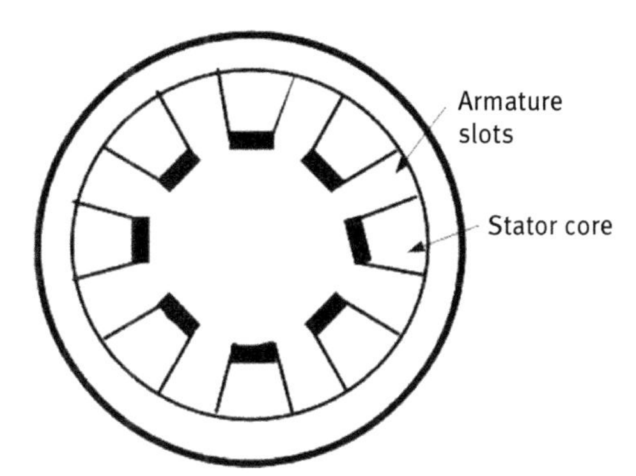

Fig. 5.1: Stator of synchronous generators.

speeds. Therefore, it has higher diameter and lower axial length to provide larger number of poles.

Cylindrical rotor

The cylindrical rotor is known as nonsalient pole rotor. The outer periphery of the rotor is constructed to form a smooth cylindrical structure and offers uniform air gap. This type of rotors have fewer poles to operate at high speeds. Therefore, it has smaller diameter and higher axial length to provide larger number of poles. The smooth structure provides less noise during operation and low winding losses.

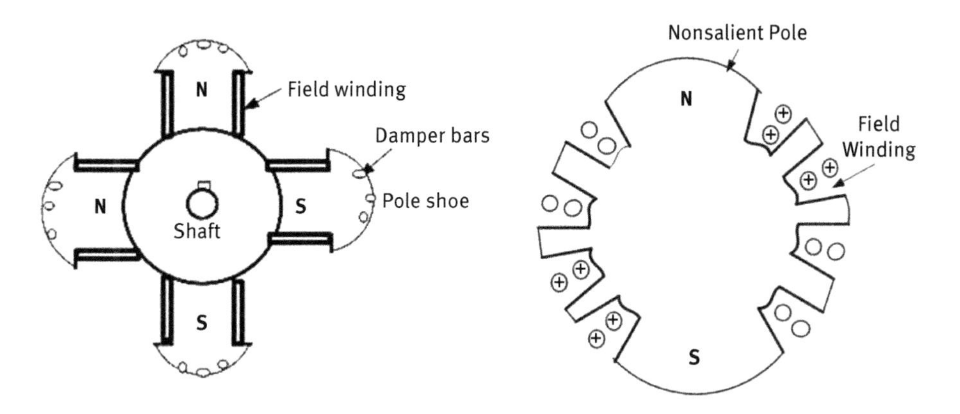

Fig. 5.2: Rotor of synchronous generators.

5.2.2 Working

The principle of working of an alternator is electromagnetic induction. Because of the relative speed between the conductors and flux, a voltage is induced in the conductors. Let us assume two opposite magnetic poles and a rectangular coil or turn placed as shown Fig. 5.3.

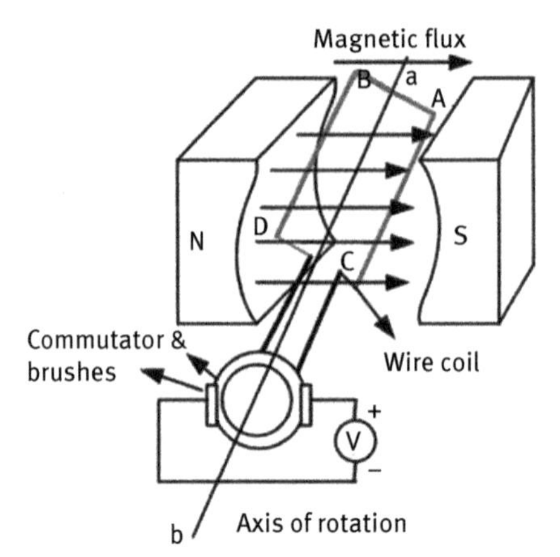

Fig. 5.3: Principle of synchronous generator.

If the armature conductor rotates in the clockwise direction, after completing 90° rotation the conductor sides AB and CD comes perpendicular to flux lines. In this position, the rate of change of flux linkage is maximum and, hence, maximum voltage is induced in the armature conductors and its direction is identified by using Fleming's left-hand rule.

Further, the conductor is rotated by 180°, 270° in a clockwise and, thus, for one revolution the conductor generates one full cycle as shown in Fig. 5.4 and is termed as the production of alternating current by revolving a turn inside a magnetic field.

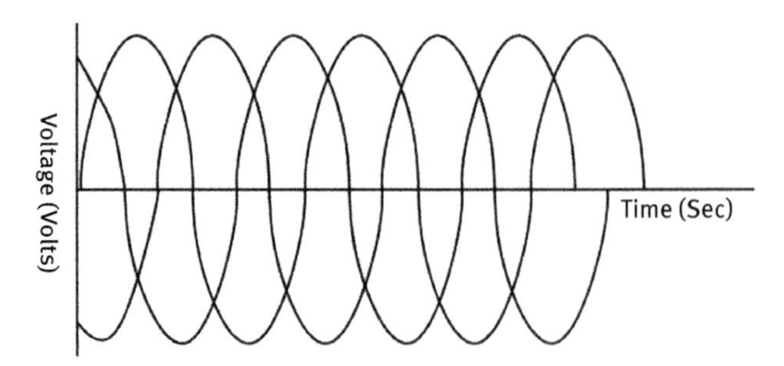

Fig. 5.4: Alternating Voltage of synchronous generator.

5.3 Emf equation of a synchronous generator

The voltage equation of a synchronous generator is derived as shown below.
Let,

P : No. of poles
ϕ : Flux/pole (Wb)
N : Speed (rpm)
f : Frequency (Hz)
Z_{ph} : No. of conductors connected in series/phase
T_{ph} : No. of turns connected in series/phase
Kc : Coil span factor
Kd : Distribution factor

Flux cut by each armature conductor in one revolution is Pϕ Weber. Time taken to complete one revolution is given by 60/N sec.

$$\text{Average emf induced per conductor} = \frac{P\Phi}{\frac{60}{N}} = \frac{P\Phi N}{60}\,\text{Volts} \tag{5.1}$$

$$\text{Average emf induced per phase} = \frac{P\phi N}{60} \times Z_{ph} = \frac{P\phi N}{60} \times 2T_{ph}$$

$$T_{ph} = \frac{Z_{ph}}{2}$$

$$\text{Average emf} = 4\,\phi\,T_{ph} = \frac{PN}{120} = 4\,\phi\,f\,T_{ph} \tag{5.2}$$

RMS value of the emf induced per phase $E_{ph} = 1.11 \times 4\,\phi\,f\,T_{ph} = 4.44\,\phi\,f\,T_{ph}$

If the coil span factor, K_c, and the distribution factor, K_d, are taken into consideration then the actual emf induced per phase is given as

$$E_{ph} = 4.44\,K_c\,K_d\phi\,f\,T_{ph} \tag{5.3}$$

Equation (5.3) is the emf equation of the synchronous generator.

Coil span factor (K_c)

It is defined as the ratio of the induced voltage in a coil when the winding is short pitched to the induced voltage in the same coil when the winding is full pitched.

Distribution factor (K_d)

It is defined as the ratio of induced voltage in the coil group when the winding is distributed in several slots to the induced voltage in the coil group when the winding is concentrated in one slot.

5.4 Voltage regulation of a synchronous generator

The voltage regulation of a synchronous generator is the difference between no-load voltage and full-load voltage expressed as a percentage of no-load voltage.

It is expressed by the equation shown below.

$$\text{Per unit voltage regulation} = \frac{|E| - |V|}{|V|} \tag{5.4}$$

$$\text{Percentage of voltage regulation} = \frac{|E| - |V|}{|V|} \times 100$$

Where,
$|E|$ = No-load voltage of the generator per phase
$|V|$ = Full-load voltage per phase

5.4.1 Determination of voltage regulation

The following methods are employed to determine the regulation of the synchronous generator.

They are:
(i) Emf or synchronous impedance method
(ii) MMF method
(iii) ZPF method
(iv) ASA method

(i) Synchronous impedance (or) emf method
In this method, to calculate the regulation of an alternator, the effect of the armature reaction is replaced by an imaginary reactance. The generator armature resistance per phase is calculated with the help of open-circuit characteristics. The full-load short-circuit current at rated voltage is calculated using short-circuit characteristics.

Voltage equation of the generator given below:

$$V = E - I_a Z_s$$

Where, $Z_s = R_a + jX_s$

Synchronous impedance is found from the OCC and SC characteristics and no-load voltage is found using equation:

$$E_o = \sqrt{(V\cos\phi + I_a R_a)^2 - (V\sin\phi + I_a X_s)^2}\,;\ + \text{for lagging},\ - \text{for leading} \tag{5.5}$$

$$\%\,\text{Regulation} = \frac{E_0 - V}{V} \times 100$$

5.4.2 Measurement of synchronous impedance

The synchronous impedance (Z_S) is measured by conducting open-circuit/no-load and short-circuit/full-load test.

Open circuit characteristics test

The alternator is made to run at its rated speed without any load, then by gradually increasing the field current of the generator the induced voltage is recorded. Field current has to be varied until 125% of the rated voltage is generated. The circuit diagram is shown in Fig. 5.5. A graph is plotted between the open-circuit phase voltage $E_p = E_t/\sqrt{3}$ and the field current I_f. The shape of this curve is same as magnetization curve. The linear portion of the OCC is extended to form an air gap line. The open circuit characteristic (OCC) and the air gap line are shown in Fig. 5.6.

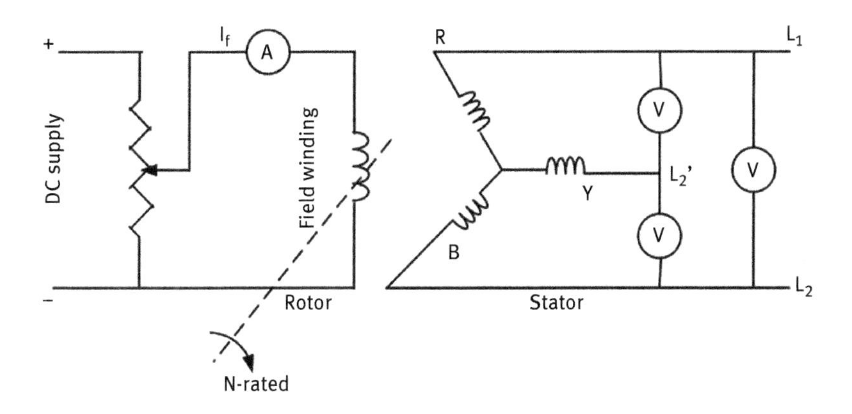

Fig. 5.5: Open circuit test of synchronous generator.

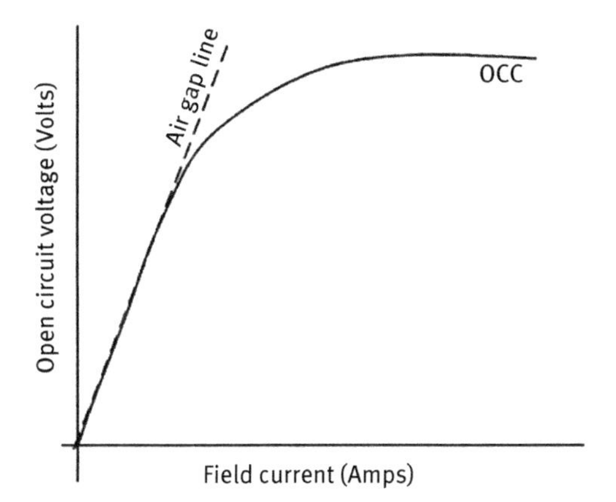

Fig. 5.6: Open circuit characteristics of synchronous generator.

Short circuit test

In the short-circuit test, the armature terminals are shorted through three ammeters as shown in Fig. 5.7. After conducting OCC test, the field current of the generator is brought to zero to make zero generated voltage. Then, the armature terminals of the generator should be short-circuited through ammeters. The generator field current is adjusted to circulate 125% to 150% of the rated value of armature current. A graph is drawn between the armature and field current I_a and I_f. This characteristic is called short circuit characteristic (SCC) and it is a straight line as shown in Fig. 5.8.

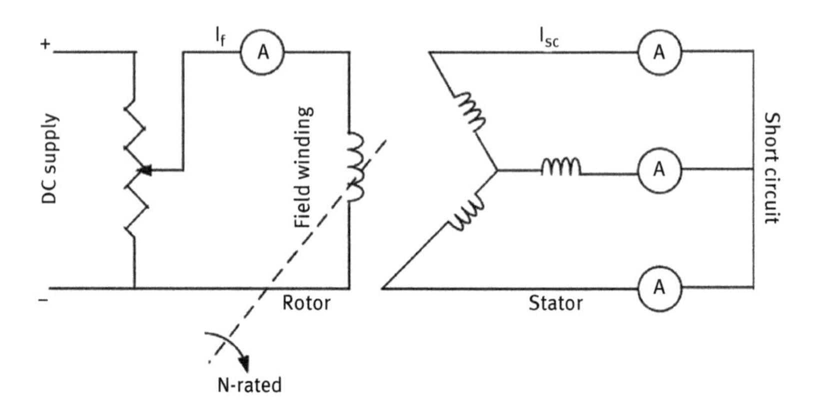

Fig. 5.7: Short circuit test of synchronous generator.

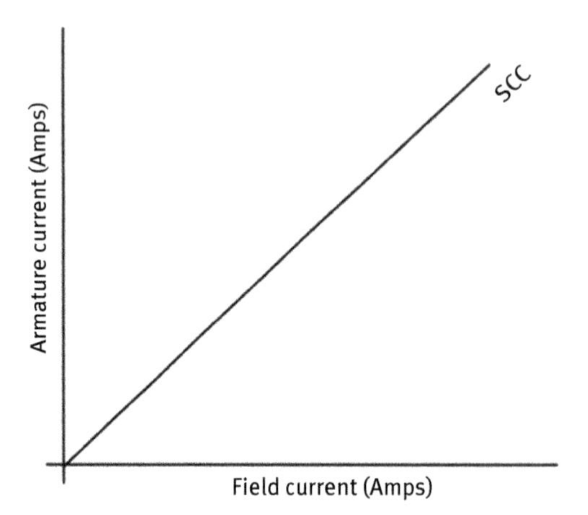

Fig. 5.8: Short circuit characteristics of synchronous generator.

Calculation of synchronous impedance

With the following procedure, the synchronous impedance is calculated.

(i) The OCC and SC characteristics are plotted on the same graph.

(ii) Measure the full-load value of short-circuit current, I_{SC}, at the rated per phase voltage of the generator.

The synchronous impedance $Z_S = \frac{\text{Open Circuit Voltage}}{\text{Short Circuit Armature Current}}$ at the same value of I_f

The synchronous reactance is determined as

$$X_s = \sqrt{Z_s^2 - R_a^2}$$

The graph is shown below.

From Fig. 5.9, field current I_f = OA which produces rated generator voltage per phase, the open-circuit voltage is AB corresponding to field current OA.

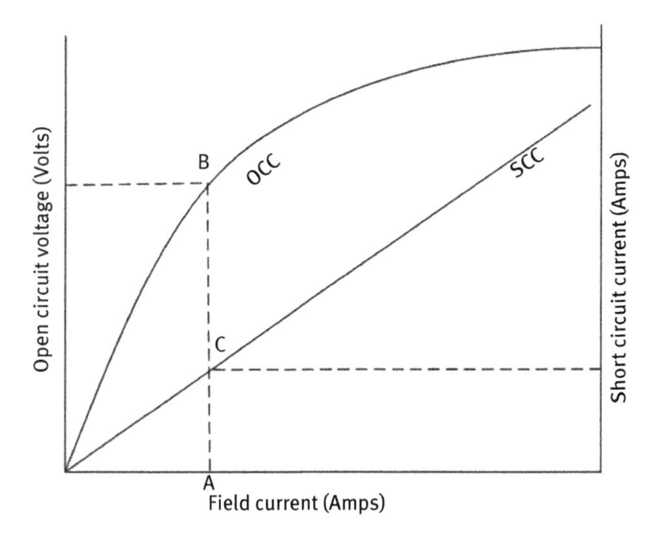

Fig. 5.9: Synchronous generator OCC and SCC characteristics to find Z_S.

Therefore,

$$Z_S = \frac{AB(\text{in Volts})}{AC(\text{ in Ampere})}$$

Assumptions in the synchronous impedance method

i) The synchronous Impedance is constant.

ii) The flux under test conditions is the same as that under load conditions.

iii) The effect of the armature reaction flux is replaced by a voltage drop proportional to the armature current and that the armature reaction voltage drop is added to the armature reactance voltage drop.
iv) The reluctance to the armature flux is constant regardless of the power factor.

5.5 Salient pole alternators and Blondel's two reaction theory

In nonsalient or smooth cylindrical type rotor, the air gap is uniform hence the reactance offered by the flux is same throughout the air gap. Therefore, this reactance is called synchronous reactance. But in salient pole type rotor the air gap is not uniform, hence the reactance is not uniform. To analyze the armature reaction, the synchronous reactance is split into two parts.

The reactance along the direct axis of the rotor is called direct axis reactance (d-axis) and the reactance perpendicular to the rotor axis is called quadrature reactance (q-axis).

The effect MMF along the d-axis will be differing with q-axis reactance, hence unequal value of the reactance is offered to the stator flux. The air gap length is very less along the d-axis, hence offered less reactance, and along the q-axis offer high reactance due to higher air gap length. Therefore, armature reaction also differs along the "d" and q-axis, where X_{ad} is direct axis reactance and X_{aq} is quadrature axis reactance.

5.6 Direct-axis and quadrature-axis synchronous reactances

The effects of the quadrature axis reactance and direct-axis reactance components of the armature reaction are explained by Blondel's two reaction theory.

By neglecting saturation, assume X_{ad} and X_{aq} of armature-reaction reactance respectively, the actual leakage reactance will be added to the armature reaction coefficients. The synchronous reactance along the direct axis is $X_{sd} = X_{ad} + X$ and synchronous reactance along the q- axis $X_{sq} = X_{aq} + X$.

In a salient–pole generator, the direct–axis reactance (X_{ad}) is higher than, the quadrature-axis reactance (X_{aq}), The phasors are clearly indicated, with reference to the phasor diagram of a (salient pole) synchronous generator supplying a lagging power factor (pf) load, shown in Fig. 5.10.

$$I_{aq} = I_a \cos(\delta + \phi)$$

$$I_{ad} = I_a \sin(\delta + \phi)$$

and $I_a = \sqrt{I_{aq}^2 + I_{ad}^2}$

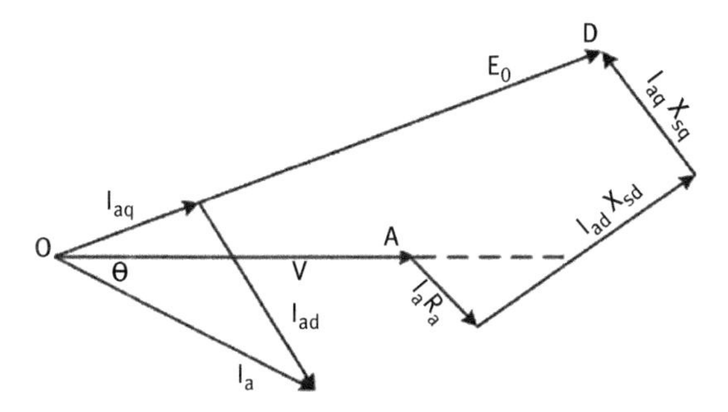

Fig. 5.10: Phasor diagram of a salient pole synchronous generator lagging pf load.

Where,
δ = Torque or power angle
ϕ = Power factor of the load

5.7 Testing of synchronous generators

5.7.1 Regulation of three-phase alternator by synchronous impedance method

The regulation of a three-phase alternator is evaluated by using synchronous impedance method.

5.7.1.1 Apparatus required

Table 5.1: List of required apparatus.

Name	Range	Quantity
Voltmeters	0–300 V MI, MC	1
Ammeter	0–3 A, MI	1
Ammeter	0–10 A, MI	1
Ammeter	0–20 A, MC	1
Ammeter	0–10 A, MI	1
Ammeter	0–2 A, MC	1
Rheostat	145 Ω/2.8 A	2
Rheostat	400 Ω/1.7 A	I

5.7.1.2 Circuit diagram

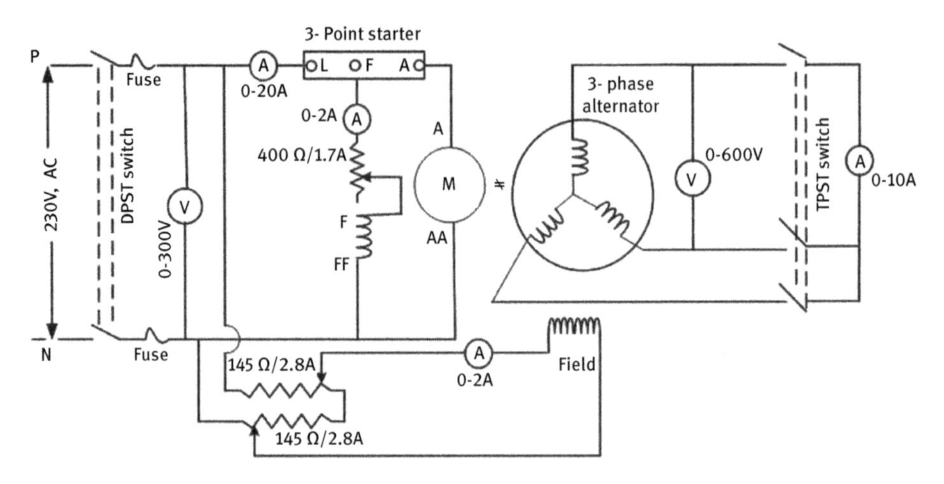

Fig. 5.11: Circuit diagram of synchronous generator: determination of regulation.

5.7.1.3 Theory

The regulation of alternator is determined at different power factors. The emf method is also known as pessimistic method as the value of regulation calculated is more than the real value. In this method, the OC and SC test data are utilized.

Synchronous impedance $Z_s = \dfrac{\text{Open Circuit Voltage}}{\text{Short Circuit Armature Current}}$ for the same field current

Synchronous reactance is determined as $X_s = \sqrt{Z_s^2 - R_a^2}$

The relationship between terminal voltage and induced e.m.f for a lagging leading power factor load is by the expression:

$$E_o = \sqrt{(V\cos\phi + I_a R_a)^2 - (V\sin\phi + I_a X_s)^2}\text{ ; + for lagging, – for leading}$$

$$\%\,\text{Regulation} = \frac{E_0 - V}{V} \times 100$$

5.7.1.4 Procedure

Open circuit test

1. The name plate details of an alternator should be noted.
2. The connections of various terminals are connected as per the circuit with DPST off position.
3. Keep the motor field rheostat (R_{fm}) in the minimum position, armature rheostat in maximum position start the motor by using three-point starter.

4. The speed is adjusted to the rated speed by adjusting the field rheostat and armature rheostat of the motor.
5. Now, switch on the generator side DPST and apply DC excitation voltage to the generator field.
6. By increasing the field current of the generator by adjusting generator field, note down the values of I_f and E_g upto 125% of the generator rated voltage.
7. The speed of the generator should be maintained constant throughout the experiment.

SC test

1. Reduce the generated voltage to "Zero" by reducing the field excitation of the generator.
2. Maintain the speed of the alternator rated speed.
3. Now close the TPST, which connects ammeters to read the short circuit current.
4. Increase the excitation of the generator until rated short circuit current is read by the ammeters.
5. Now gradually reduce the excitation and open the TPST.
6. Draw the magnetization plot between E_g verses I_f, draw the SC characteristics and calculate Z_s and X_s.

5.7.1.5 Observation table

OC test

Table 5.2: Observation table.

S.No.	Field current, I_f (A)	Generated emf, E_0 (V)
1		

SC test

Table 5.3: Observation table.

S.No.	Field current, I_f (A)	Short-circuit current, I_{sc} (A)
1		

5.7.1.6 Model graph

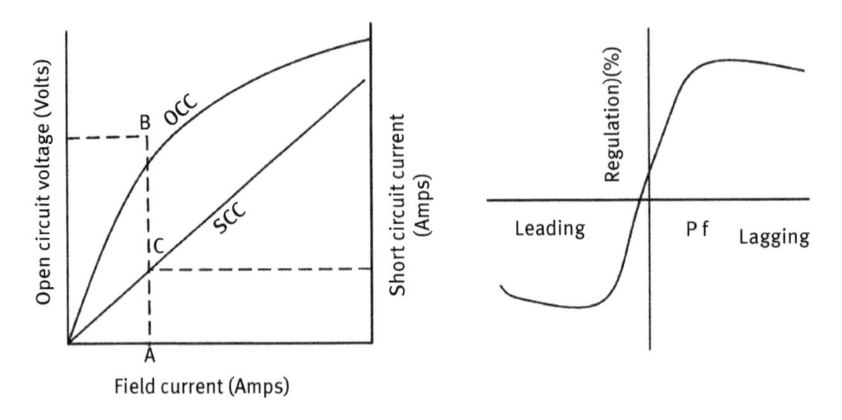

Fig. 5.12: Model graph of characteristics of synchronous generator.

5.7.1.7 Practical calculations

The regulation of the alternator test with the following rating of a machine and the data is recorded.

DC motor	Alternator
O/p power = 5.2 kW	Rating = 5 kVA
Voltage = 220 V	Excitation = 7 A, 240 V
Current = 29 A	Voltage = 415 V
speed = 1,500 rpm	Current = 7 A

Observation table
OC test

Table 5.4: Practical observations.

S No.	Field current, I_f (A)	Generated emf, E_0 (V)
1	0.2	80
2	0.35	121
3	0.5	174
4	0.8	200
5	0.9	244
6	1.1	270
7	1.6	271

SC test

Table 5.5: Practical observations.

S.No.	Field current, I_f (A)	Short-circuit current, I_{sc}(A)
1	0.45	7

For the rated voltage (V_{OC}) = 230 V

Short-circuit current (I_{SC}) = 12.7 A

Armature resistance (R_a) = 1.8 Ω

From the Graph,

Synchronous impedance $(Z_{sc}) = \frac{V_{OC}}{I_{SC}} = \frac{230}{13.4} = 17.16\Omega$

$$X_s = \sqrt{Z_s{}^2 - R_a{}^2} = \sqrt{17.16^2 - 1.8^2} = 17.04\Omega$$

Line voltage (V_{line}) = 415 V

Phase voltage $(V_{phase}) = \frac{415}{\sqrt{3}} = 239.6\ V$

The relation between terminal voltage and induced emf with leading and lagging power factor is given by

$$E = \sqrt{(V_{ph}\cos\phi + I_aR_a)^2 \overset{+}{-} (V_{ph}\sin\phi + I_aX_s)^2}\ \ + \text{for leading and} - \text{for lagging}$$

i) At $\cos\phi = 1, \sin\phi = 0,\ I_a = 7$

$$\text{Induced E.M.FE} = \sqrt{239.60 \times 1 + 7 \times 1.8)^2 + (239.60 \times 0 + 7 \times 17.04)^2} = 278.98\ V$$

For leading = 278.98 V

ii) At $\cos\phi = 0.8, \sin\phi = 0.6$

$$E = \sqrt{(239.60 \times 0.8 + 7 \times 1.8)^2 + (239.60 \times 0.6 + 7 \times 17.04)^2} = 333.04\ V$$

For leading = 175.6 V

iii) At $\cos\phi = 0.6, \sin\phi = 0.8$

$$E = \sqrt{(239.60 \times 0.6 + 7 \times 1.8)^2 + (239.60 \times 0.8 + 7 \times 17.04)^2} = 348.05\ V$$

For leading = 138.6 V

$$\text{Percentage Voltage Regulation} = \frac{E - E_{ph}}{E_{ph}} \times 100$$

i) At $E = 218.99\,V,\ E_{ph} = 239.60\,V$

$$\%\text{Regulation} = \frac{278.99 - 239.69}{239.60} \times 100 = 16.43\% \text{ for lag}$$

ii) At $E = 333.04\,V,\ E_{ph} = 239.06\,V$

$$\%\text{Regulation} = \frac{333.04 - 239.60}{239.60} \times 100 = 38.99\% \text{ lag}$$

$$= -47.8\% \text{ lead}$$

iii) At $E = 348.05\,V,\ E_{ph} = 239.06\,V$

$$\%\text{Regulation} = \frac{348.05 - 239.60}{239.60} \times 100 = 45.26\% \text{ lag}$$

$$= -42.25\% \text{ lead}$$

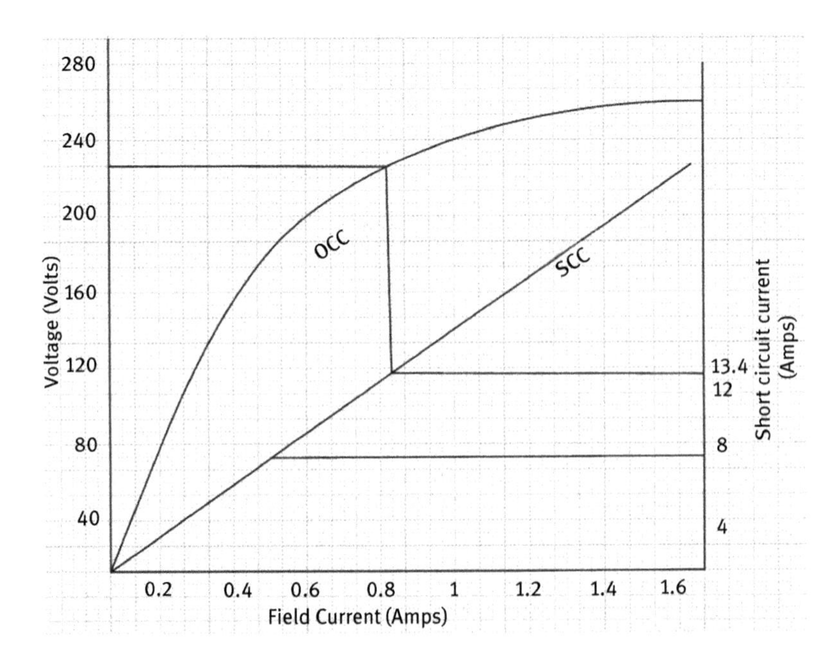

Fig. 5.13: Graph of characteristics of synchronous generator.

5.7.1.8 Result

Regulation of alternator is determined by synchronous impudence method and graph is drawn at various power factors.

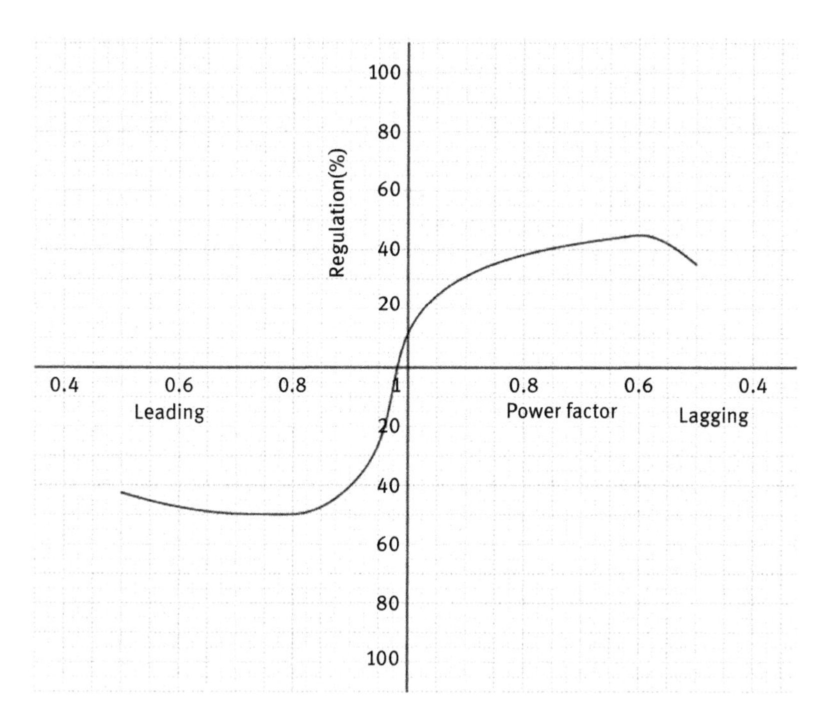

Fig. 5.14: Graph of pf vs. regulation of synchronous generator.

5.7.2 Slip test

The direct and quadrature axis reactance (X_d and X_q) are evaluated by performing a slip test on a salient pole synchronous machine.

5.7.2.1 Apparatus required

Table 5.6: List of required apparatus.

Name	Range	Quantity
Voltmeter	0–300 V, MI	1
Voltmeter	0–300 V, MC	1
Ammeter	0–2 A, MC	1
Ammeter	0–10 A, MI	1
three-phase variac	440/0–470 V, 12 A	1
Rheostat	400 Ω/1.7 A	1

Table 5.6 (continued)

Name	Range	Quantity
Phase sequence indicator		1
Digital tachometer		1
Connecting wires		

5.7.2.2 Circuit diagram

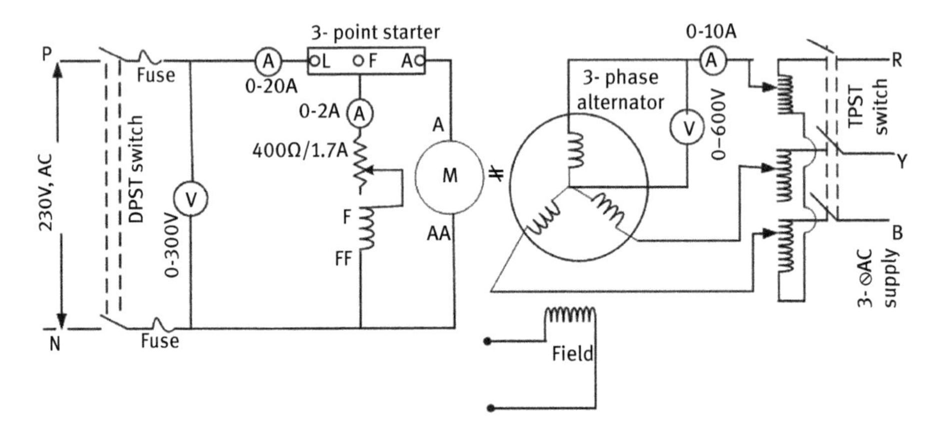

Fig. 5.15: Circuit diagram of synchronous generator: slip test.

5.7.2.3 Theory

In nonsalient or smooth cylindrical type rotor, the air gap is uniform and hence the reactance offered by the flux is same in the air gap. Therefore, this reactance is called synchronous reactance. But in salient pole type rotor the air gap is not uniform hence the reactance is not uniform. To analyze the armature reaction, the synchronous reactance is split into two parts.

The reactance along the direct axis of the rotor is called direct axis reactance (X_d) and the reactance perpendicular to the rotor axis is called quadrature reactance (Xq). The per unit value of the X_d and X_q are calculated in laboratory using slip test.

In this test, the alternator is driven slightly lower than synchronous speed and the generated voltage and phase sequence is observed. Now, make sure that the generator voltage phase sequence and three-phase external supply phase sequence are same before applying this voltage to the stator of alternator.

Now, open its field winding and apply the three-phase AC voltage to the stator of alternator. Under this situation, the stator of alternator, acts as field and rotor conductor, is cut by the stator field and a voltage is induced in the rotor conductors.

As the rotor of alternator salient pole type, the induced voltage is changing between maximum and minimum position due to the unequal air gap of the rotor. Along the direct axis of the rotor maximum voltage is induced and draws minimum current as it offers lower reluctance and maximum flux linkage, and along the quadrature axis minimum voltage is induced and draws maximum current as it offers larger reluctance and minimum flux linkage.

Thus, the estimated values are given by

$$\text{Direct – axis synchronous reactance } X_d = \frac{E_{max}}{I_{min}}$$

$$\text{Quadrature – axis synchronous reactance } X_q = \frac{E_{min}}{I_{max}}$$

5.7.2.4 Procedure

1. The name plate details of the alternator and motor should be noted.
2. The connections of various terminals are connected as per the circuit with DPST off position.
3. Keep the motor field rheostat (R_{fm}) in the minimum position, armature rheostat in maximum position start the motor by using three-point starter.
4. The speed is adjusted to slightly less than the synchronous speed by adjusting the field rheostat and armature rheostat of the motor.
5. Check the phase sequence of the induced voltages of the alternator by exciting the field winding of the alternator. After checking the phase sequence open the field circuit of the alternator.
6. Now close the TPST and apply the three-phase supply to the stator of the alternator.
7. Record the maximum and minimum voltages induced in the rotor using voltmeters and also currents using ammeter.
8. Open the TPST and bring the alternator speed to minimum and switch off the panel.
9. Calculate the X_d and X_q.

5.7.2.5 Observation table

Table 5.7: Observation table.

S. No.	Maximum armature current (I,max)	Minimum armature current (I,min)	Maximum armature voltage (V,max)	Minimum armature voltage (V,min)
1				

5.7.2.6 Model calculations

$$X_d = \frac{\text{Maximum armature terminal voltage per phase}}{\text{Minimum armature current per phase}}$$

$$X_q = \frac{\text{Minimum armature terminal voltage per phase}}{\text{Maximum armature current per phase}}$$

5.7.2.7 Practical calculations

Slip test is conducted on the following rating of the motor and readings were tabulated.

Table 5.8: Machine ratings.

Motor	Generator
Rating = 3.5 kW, 1,500 rpm	Rating = 5 kVA
Voltage = 230 V	Voltage = 415 V
Load Current = 27.8 A	Current = 7 A

5.7.2.8 Observation table

Table 5.9: Practical observations.

S. No	Maximum armature current (I,max)	Minimum armature current (I,min)	Maximum armature voltage (V,max)	Minimum armature voltage (V,min)
1	4.7 amps	3.7 amps	27 volts	23 volts

$$\text{Direct axis reactance } X_d = \frac{E_{max}}{I_{min}} = \frac{27}{3.7} = 7.297\Omega$$

$$\text{Quadrature axis reactance } X_q = \frac{E_{min}}{I_{max}} = \frac{23}{4.7} = 4.89\Omega$$

5.7.2.9 Result

The direct and quadrature axis reactance (X_d and X_q) of synchronous generator is obtained by conducting slip test and their values are tabulated.

5.8 Viva-voce questions

1. Define the regulation of alternator.
2. Under what condition, regulation is positive or negative?
3. What is regulation at UPF?
4. Why regulation is so important in alternator?
5. How is regulation effected by the armature reaction?
6. Write the equation of emf induced in an alternator.
7. Classify the alternators.
8. Mention the applications of synchronous generators.
9. What are the merits of salient pole rotors?
10. Why is laminated core used in an alternator?
11. Define efficiency of an alternator.
12. When alternator is loaded, the voltage changes. Why?
13. Define infinite bus.
14. Define slip.
15. Define direct axis synchronous reactance.
16. Define quadrature axis synchronous reactance.
17. Explain the method of conducting slip test?
18. How do you find the regulation of synchronous generator?
19. How do you calculate synchronous impedance (Z_s) using emf method?
20. Why regulation calculated using emf method is pessimistic. Why?
21. Explain two reaction theory.
22. Why is regulation calculated using MMF method optimistic. Why?

5.9 Objective questions

1. Using the ZPF method —— of an alternator is found.
   (a) X_q (b) X_d (c) slip (d) regulation

2. As the speed of an synchronous generator increases, the frequency ——
   (a) increases (b) decreases (c) remains constant (d) none

3. In a 50 kVA, 440V, 50Hz, star connected alternator, the effective armature resistance is 0.25 Ω/phase. The synchronous reactance is 3.2 Ω/phase and leakage reactance is 0.5 Ω/phase, full-load output current at unity power factor will be
   (a) 65.6 A (b) 25.4 A (c) 75.6 A (d) 5.4 A

4. In the above problem, full-load line voltage will be
   (a) 3,030 V (b) 471 V (c) 353 V (d) 336 V

5. In salient pole machines, the armature MMF cannot be accounted for by introducing one equivalent reactance due to
   (a) nonuniform air gap
   (b) variable resistance
   (c) uniform air gap
   (d) all

6. If the internal power factor angle of synchronous motor is 30 degree. Then the direct axis component of the armature current will be
   (a) 0.5 pu (b) 0.567 pu (c) 5.43 pu (d) 5.01 pu

7. If the internal pf angle of a synchronous motor is 30 degree. Then the quadrature axis component of the armature is
   (a) 0.866 pu (b) 0.25 pu (c) 3.03 pu (d) 4.04 pu

8. For a 400 V, three-phase alternator gives an open-circuit voltage of 380 V and armature current of 38 A at a field current of 20 A. Then the synchronous reactance of the machines
   (a) 21.4 ohm (b) $81/\sqrt{2}$ ohm (c) 91 ohm (d) 10 ohm

9. For obtaining maximum current when we conduct the "Slip Test" on a synchronous machine, its armature field will align along
   (a) k-axis (b) q-axis (c) d-axis (d) p-axis

10. Stability of a synchronous motor ——— with the increase in excitation.
    (a) increases (b) decreases (c) unaffected (d) none

11. Synchronizing power of a synchronous motor varies
    (a) directly as synchronous reactance X_s
    (b) directly as the square of the synchronous reactance, X_s^2
    (c) inversely as $1/X_s$
    (d) all

Answers

1. d, 2. a, 3. a, 4. b, 5. a, 6. a, 7. a, 8. d, 9. b, 10. a, 11. c

5.10 Exercise problems

1. The phase voltage of a three-phase, 50 Hz alternator consists of a fundamental, a 20 % third harmonic and a 10% fifth harmonic. The amplitude of the fundamental voltage is 1,000 V. Calculate the RMS line voltage when the alternator windings are in (i) Star and (ii) Delta.

2. A 220 V, 50 Hz, four-pole star connected alternator with resistance of 0.06 Ω/phase are the following data for open-circuit and full-load characteristics:

Field current(A)	0.2	0.4	0.6	0.8	1.00	1.2	1.4	1.8	2.2	2.6	3.0	3.4
Open-circuit Voltage (volts)	29.0	58.0	87.0	116	146	172	194	232	261.5	284	300	310

Find the percentage voltage regulation at a full-load current of 40 amps at a power factor of 0.8 lagging using synchronous impedance method.

A 500 kVA, 1,100 V, 50 Hz star connected 3-phase alternator has armature resistance per phase of 0.1 Ω and synchronous reactance per phase of 1.5 Ω. Find its voltage for (i) 0.9 pf lag and (ii) 0.8 pf lead. Also, find the voltage regulation in each case.

6 Synchronous Motor

6.1 Introduction

The synchronous motor similar to all motors converts electrical energy into mechanical energy running at synchronous speed. Its speed depends up on the frequency of the supply voltage and number of poles. This machine is a doubly excited machine where its stator is connected to three-phase supply and rotor is excited with DC supply. It works on the principle of magnetic locking. These motors are used as condensers for large power systems to correct the power factors, because the reactive power produced by a synchronous motor can be controlled by controlling the field excitation. The synchronous condensers are more reasonable in the large sizes than static capacitors.

6.2 Construction and working of synchronous motor

The synchronous motor consists of stator and a rotor.

1. Stator

The stator is made with cast iron for low rating machines and for high rating welded steel is used. Stator core is laminated to minimize core losses and it is made of silicon steel. A three-phase star (Y) connected winding is wounded in the slots and its neutral is connected to the ground

2. Rotor

The rotor is wounded with a field winding and is connected to DC through two slip rings by a separate DC source.

Rotor construction is of two types, namely:

(i) Salient pole type. (ii) Non-salient or cylindrical pole type.

i. Salient pole type

In salient pole type rotor the poles are projecting outside from the core of the rotor surface and offers nonuniform air gap as shown in Fig. 6.1. These types of rotors are employing concentric winding and damper bars are in the rotor pole faces to damp out the rotor oscillations during the rapid changes in the load. The rotor pole faces are constructed to produce a sinusoidal voltage. This type of rotors has many poles to operate at low speeds. Therefore, it has higher diameter and lower axial length to provide larger number of poles.

https://doi.org/10.1515/9783110682274-006

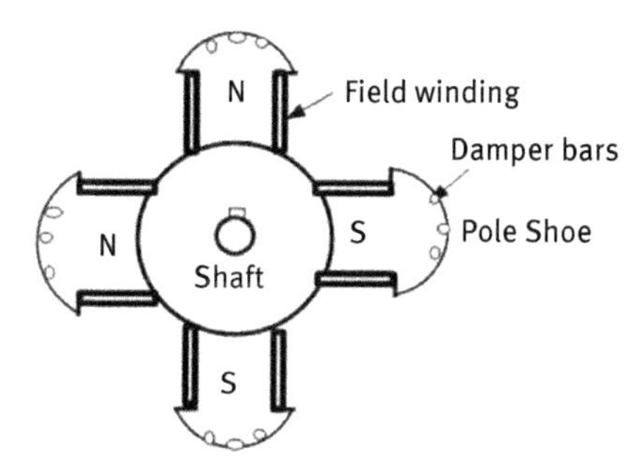

Fig. 6.1: Synchronous motor – salient pole type.

ii. Cylindrical rotor or non-salient pole

The cylindrical rotor is known as nonsalient pole rotor. The outer periphery of the rotor is constructed to form a smooth cylindrical structure and offers uniform air as shown in Fig. 6.2. This type of rotors has fewer poles to operate at high speeds. Therefore, it has smaller diameter and higher axial length to provide larger number of poles. The smooth structure provides less noise during operation and low winding losses. This type of construction has mechanical robustness and gives noiseless operation at high speeds.

Since steam turbines run at high speed and a frequency of 50 Hz needs a small number of poles on the rotor of high-speed alternators (also called turbo-alternators).

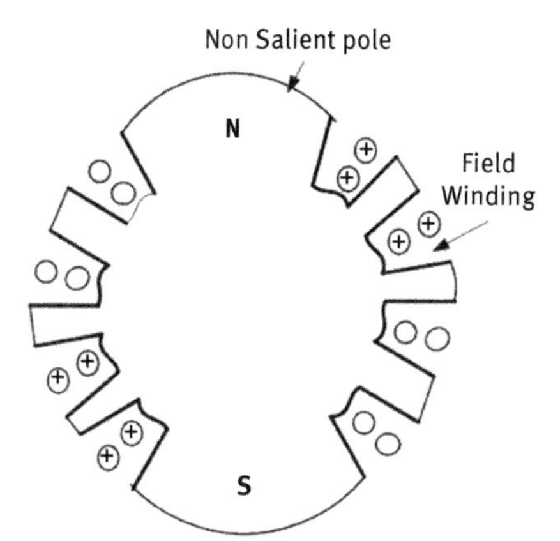

Fig. 6.2: Synchronous motor– non–salient pole (cylindrical) type.

The stator is wound for a similar number of poles as that of the rotor and fed with three-phase AC supply. The three-phase AC supply produces a rotating magnetic field in the stator. The rotor winding is fed with DC supply which magnetizes the rotor.

Working of synchronous motor

Consider a two-pole synchronous motor as shown in Fig. 6.3.

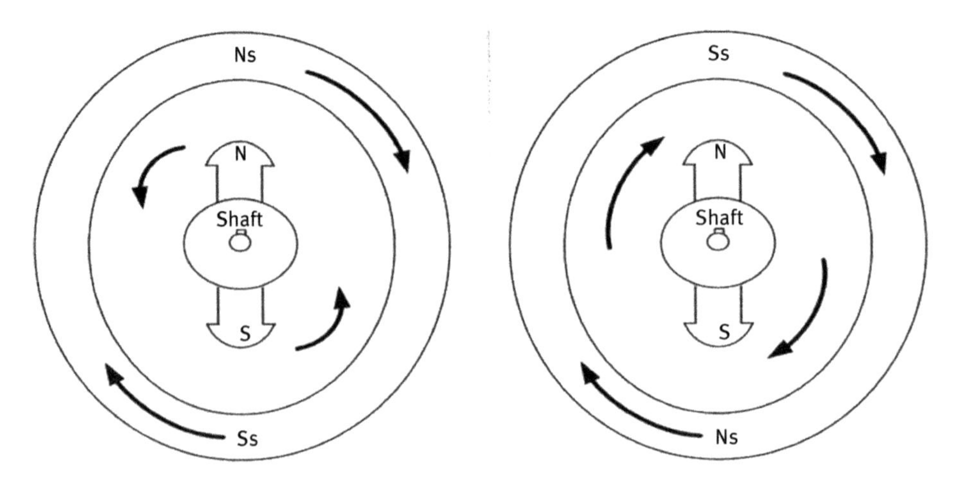

Fig. 6.3: Synchronous motor – two-pole synchronous machine.

Now, the stator poles are revolving with synchronous speed (let's say clockwise). If the rotor position is such that, N pole of the rotor is near the N pole of the stator, as shown in first schematic of Fig. 6.3, then the poles of the stator and rotor will repel each other, and the torque produced will be anticlockwise.

Let us assume that a three-phase supply is connected to stator and produces rotating flux having north and south poles as shown in Fig. 6.3, which is revolving synchronous speed. As the stator and rotor poles are opposing each other and a force will be produced and rotor starts to rotate in clock wise direction. During the half cycle, the position of north and south poles are interchanged. Hence the force also changes its direction. Therefore, the net force acting on the rotor is zero and the motor does not start.

But by using external force or prime mover if the rotor is rotated at synchronous speed, the stator poles keep attracting the rotor poles till it gets locked. This is called magnetic locking. Now the rotor rotates with the stator flux speed, that is, synchronous speed in the same direction as the stator field is rotating.

6.3 Methods of starting synchronous motor

The following methods are employed to start the synchronous motor:

i. Using External prime mover: An external prime mover to rotate the rotor of synchronous motor close to its synchronous speed. This method is used in laboratory. Initially the machine is started as a generator and is then connected to three-phase supply by synchronization procedure. After the synchronization the power supply to the prime mover is disconnected so that the synchronous generator will continue to run as a synchronous motor.

ii. Using damper windings: The damper windings are inserted in the rotor poles. Initially the synchronous motor runs as induction motor and as the rotor is running, the poles get locked then the motor starts running at synchronous speed.

6.4 Behavior of a synchronous motor

The behavior of a synchronous motor can be understood by considering its equivalent circuit and explained below.

6.4.1 Equivalent circuit of three-phase synchronous motor

The equivalent circuit of a nonsalient pole synchronous motor is shown in Fig. 6.4. All values are given per phase. Applying Kirchhoff's voltage law

$$V_T = I_a R_a + j I_a (X_l + X_{as}) + E_f \tag{6.1}$$

$$Xs = X_l + X_{as}$$

$$V_T = I_a R_a + jI_a Xs + E_f = I_a(R_a + jXs) + E_f = I_a Z_s + E_f \tag{6.2}$$

Where
V_T = Terminal voltage per phase volts
I_a = Armature current per phase amps
R_a = Armature resistance per phase(Ω)
X_l = Armature leakage reactance per Phase(Ω)
Xs = Synchronous reactance per phase (Ω)
Z_s = Synchronous Impedance per phase (Ω)

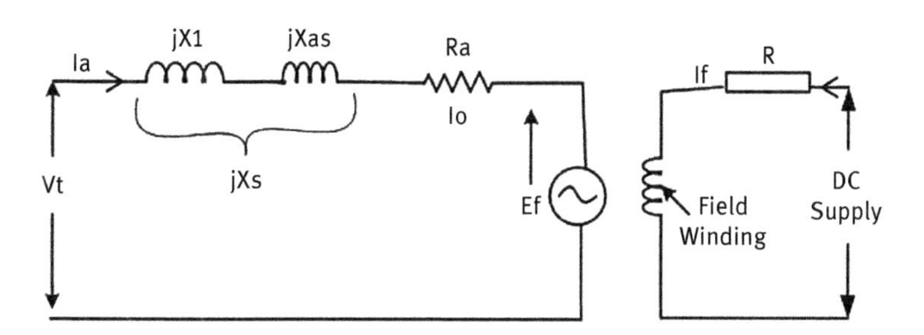

Fig. 6.4: Equivalent circuit of a synchronous motor armature.

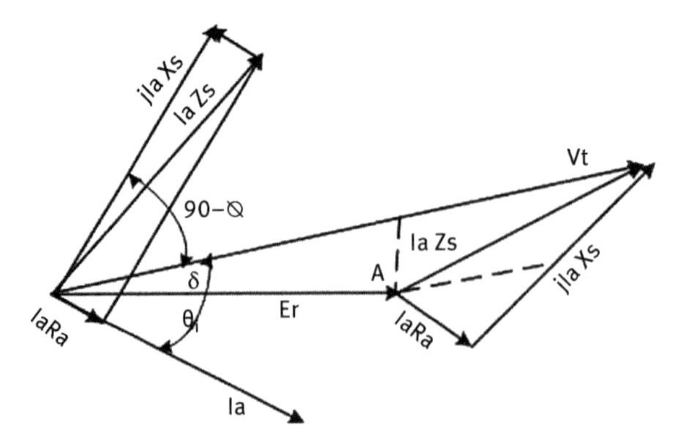

Fig. 6.5: Synchronous motor phasor diagram.

From the phasor, the emf equation

$$E_f = V_T - I_a Z_s \tag{6.3}$$

$$Zs = Ra + jXs$$

The phase angle δ between the terminal voltage V_T and the excitation voltage E_f in Fig. 6.5 is called the torque angle or load angle or power angle.

6.4.2 Synchronous-motor power equation

The equivalent circuit and phasor diagram of a synchronous motor is shown in Fig. 6.6. The resistance of armature R_a is small as compared to synchronous reactance. Hence R_a is neglected, so that eq. (6.2) can be written as $V_T = jI_a X_s + E_f$

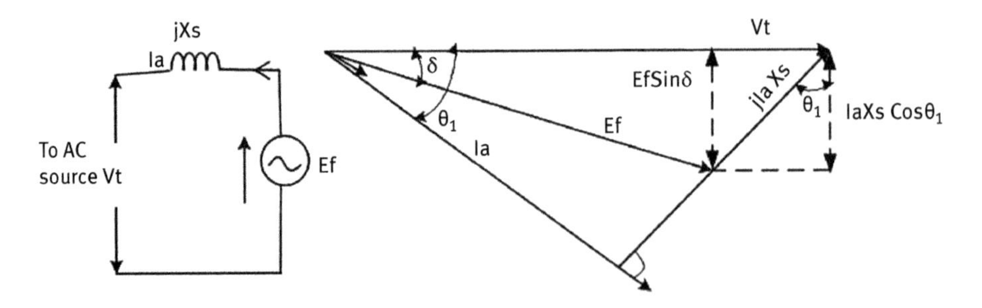

Fig. 6.6: Equivalent circuit and phasor diagram of a synchronous motor (by assuming *Ra* is negligible).

From the above phasor diagram, we can write,

$$I_a X_s \cos\phi_1 = -E_f \sin\delta$$

Multiplying by V_T on both sides

$$V_T I_a \cos\phi_1 = -\frac{V_T E_f \sin\delta}{X_s} \tag{6.4}$$

The term V_T I_a $\cos\phi_l$ is the expression for power input and it also represent the electromagnetic power developed, per phase.

Thus,

$$\text{Active power input per phase}(P_{in}) = V_T\, I_a \cos\phi_1 \tag{6.5}$$

or

$$\text{Active power input per phase}(P_{in}) = \frac{V_T E_f \text{Sin}\delta}{X_s} \tag{6.6}$$

$$\text{For a three-phase} = 3V_T I_a \cos\phi_1 = 3\frac{V_T E_f \sin\delta}{X_s} \tag{6.7}$$

Equation (6.7) is called power equation of the synchronous-machine. Assuming a constant voltage source and supply frequency, eqs. (6.5) and (6.6) may be written as

$$P = E_f \sin\delta \tag{6.8}$$

$$P = I_a \cos\phi_1 \tag{6.9}$$

The eqs. (6.8) and (6.9) are very helpful in analyzing the synchronous-motor behavior.

6.5 Effect of changes in field excitation on synchronous motor performance

If the shaft load is assumed constant, the steady-state value of E_f $\sin\delta$ should be also be constant.

Any raise in E_f will cause a transient increase in E_f $\sin\delta$, and the rotor will speed up. As the rotor changes its angular position, δ decreases until E_f $\sin\delta$ has the same steady-state value as before, at which time the rotor is again operating at synchronous speed, as it should run only at the synchronous speed. This change in the angular position of the rotor magnets relative to the poles of rotating the magnetic field of the stator occurs in a fraction of a second. The effect of changes in field excitation on armature current, power angle and power factor of a synchronous motor operating with a constant shaft load, from a constant voltage, constant frequency supply, is illustrated in Fig. 6.7.

From eq. (6.8) we have for a constant shaft load,

$E_{f1} \sin\delta = E_{f2} \sin\delta = E_{f3} \sin\delta = E_f \sin\delta$

This is shown in Fig. 6.8, where the locus of the tip of the E_f phasor is a straight line parallel to

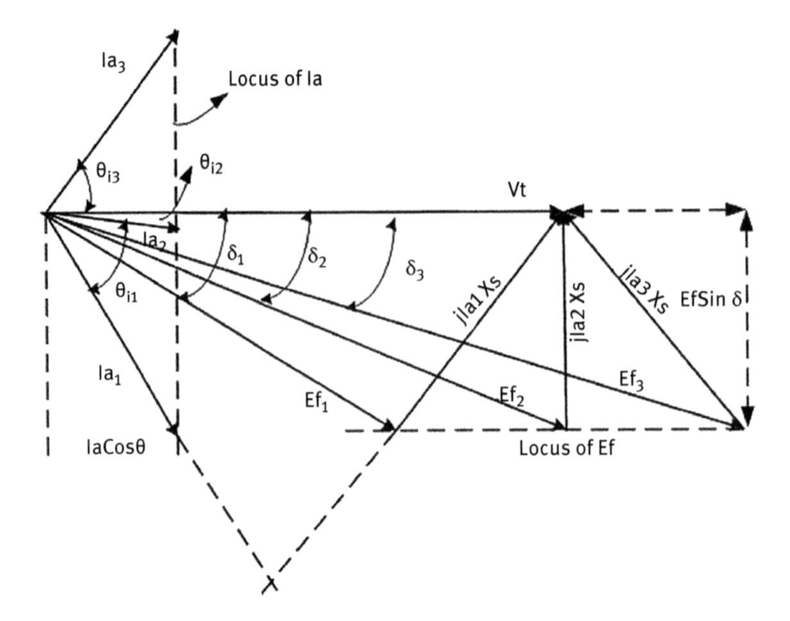

Fig. 6.7: Phasor diagram showing the effect of changes in field excitation on armature current, power angle and power factor of a synchronous motor.

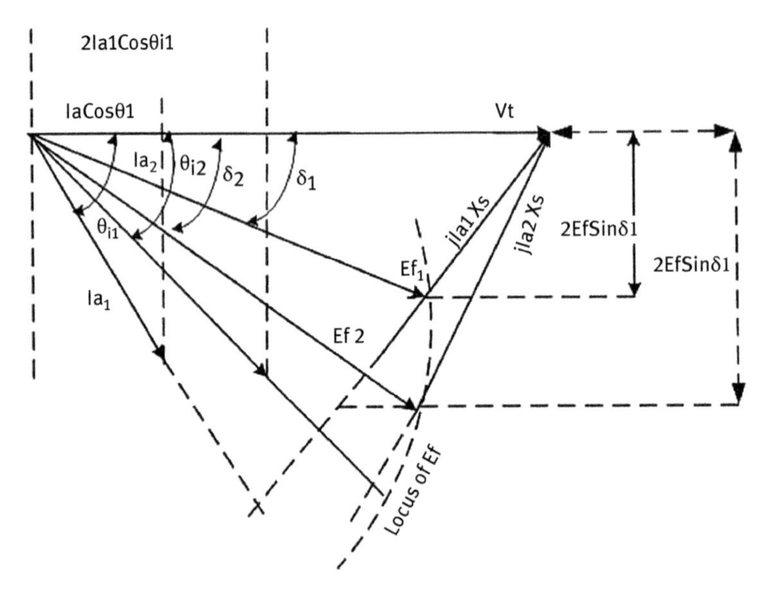

Fig. 6.8: Phasor diagram showing the effect of changes in shaft load on armature current, power angle and power factor of a synchronous motor.

the V_T phasor. Similarly, from eq. (6.9) for a constant shaft load,

$I_{a1} \cos\phi_l = I_{a2} \cos\phi_l = I_{a2} \cos\phi_l = I_a \cos\phi_l$

This is also shown in Fig. 6.7 where the locus of the tip of the I_a phasor is a line perpendicular to the V_T phasor.

Note that increasing the excitation from E_{f1} to E_{f3} in Fig. 6.7 caused the phase angle of the current phasor with respect to the terminal voltage V_T (and hence the power factor) to go from lagging to leading. The value of field excitation that results in unity power factor is called normal excitation. Excitation greater than normal is called over excitation, and excitation less than normal is called under excitation. Furthermore, as indicated in Fig. 6.8, when operating in the overexcited mode, $|E_f| > |V_T|$. In fact, a synchronous motor operating under overexcitation condition is sometimes called a synchronous condenser.

6.6 V and inverted V curve of a synchronous motor

The V curve is drawn between the armature current I_a and field current I_f at no-load. Since the plot shape is similar to the alphabet "V," therefore the curves are called V curve of synchronous motor.

The armature current of the synchronous motor decreases with increasing the field current until it reaches the minimum value. At this minimum value of armature current, the synchronous motor is said to be operate at unity power factor. The motor operates in lagging power factor before reaching this minimum point.

Further, if the field current of the motor is increased, the armature current starts increasing and the motor is said to operate in leading power factor. If this process is repeated for various increased loads, a family of graphs is obtained. The V curves of a synchronous motor are shown in Fig. 6.9 and are useful in adjusting the field current of the synchronous motor.

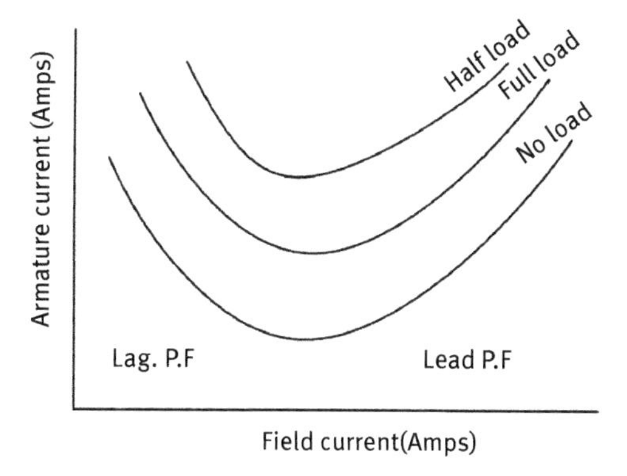

Fig. 6.9: "V" curves of synchronous motor.

Figure 6.10 below shows the graph between power factor and field current at the different loads.

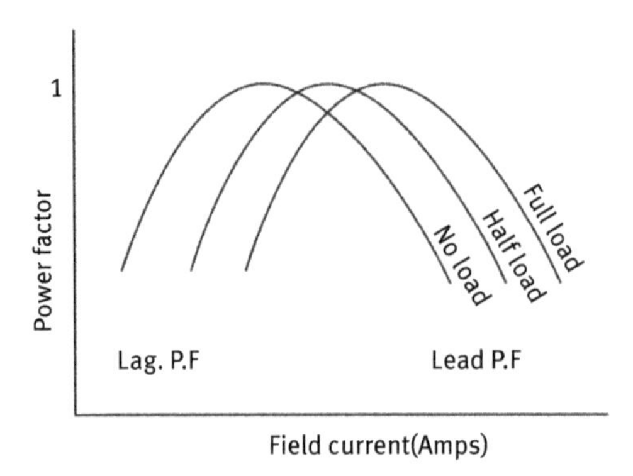

Fig. 6.10: "Inverted V" curves of synchronous motor.

It is clear from Fig. 6.10 that, if the synchronous motor at full load is operating at unity power factor, then removal of the shaft load causes the motor to operate at a leading power factor.

6.7 Testing of synchronous motor

6.7.1 "V" and inverted "V" curves of a three-phase synchronous motor

The "V" and "inverted V" curves of synchronous motor are evaluated by using the below method

6.7.1.1 Apparatus required

Table 6.1: List of required apparatus.

Name	Range	Quantity
Voltmeter	0–600 V, MI	1
Ammeter	0–10 A, MI	1
Ammeter	0–10 A, MC	1
Rheostat	400 Ω/1.7 A	1
Rheostat	40 Ω/9 A	1
Wattmeter	(600 V,10 A, UPF)	2
3-phase auto T/f	(415/0–470 V) dimmer	1

6.7.1.2 Circuit diagram

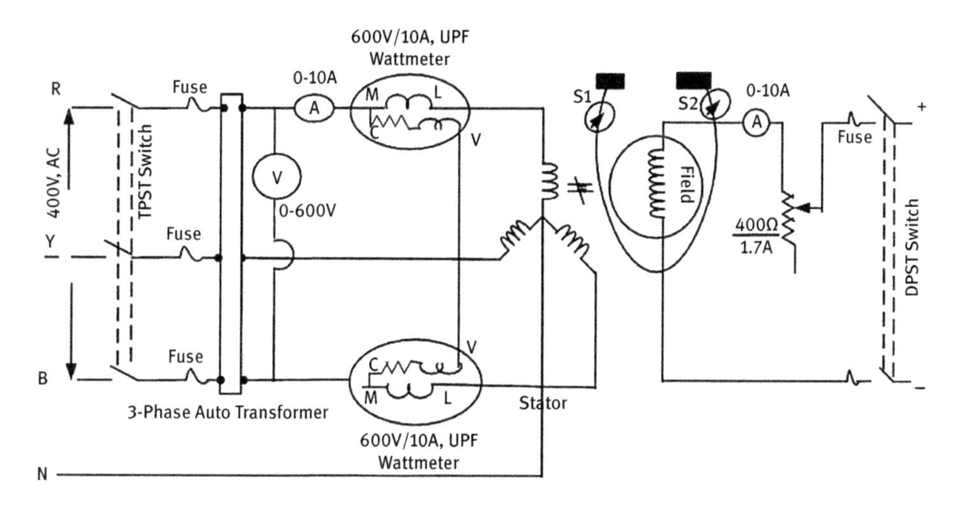

Fig. 6.11: Circuit diagram of three-phase synchronous motor: determination of “V” and inverted “V” curves.

6.7.1.3 Theory

The synchronous motor is a doubly excited motor. Its field winding is excited by separate DC source and stator is connected to three-phase supply. The field excitation affects the operating mode of the motor.

If the synchronous motor is operating under variable field excitation with a constant load, the following points were observed:

i) Under low excitation current the motor draws higher armature current.
ii) When the field excitation increases, the armature current starts reducing.

Procedure

1. The name plate details of the induction motor should be noted.
2. The connections of various terminals are connected as per the circuit with DPST off position.
3. Initially, DPST is open, P.D is kept at zero volts output position, belt around the brake drum is loosened and then TPST is closed.
4. Gradually the output voltage of 3-phase variac is increased up to rated voltage.
5. Then DPST is closed and rotor excitation is given by increasing the output of the potential divider then the rotor is pulled into synchronism and the motor runs at synchronous speed N_s as synchronous motor.
6. Then by varying PD, that is, by varying I_f from minimum to maximum different readings of I_f I_a & $\cos\varphi$ are noted at no-load.

7. Then I_a(on Y-axis) versus I_f (on X-axis) gives V curves and pf (Cosϕ)(on Y-axis) versus I_f (on X-axis) gives inverted V curves.
8. Then I_f and V_a is brought to minimum DPST and TPST are opened.
9. The same above procedure can be represented by applying a constant mechanical load (by tightening the belt around the brake drum and maintaining the constant load) for another set of V and inverted V curves.

6.7.1.4 Observation table

Table 6.2: Observation table.

S. No.	Supply voltage (V)	Wattmeter reading (W_1)	Wattmeter reading (W_2)	Field current I_f(Amp)	Armature current I_a(Amp)	Cosϕ
1						
2						

6.7.1.5 Model graph

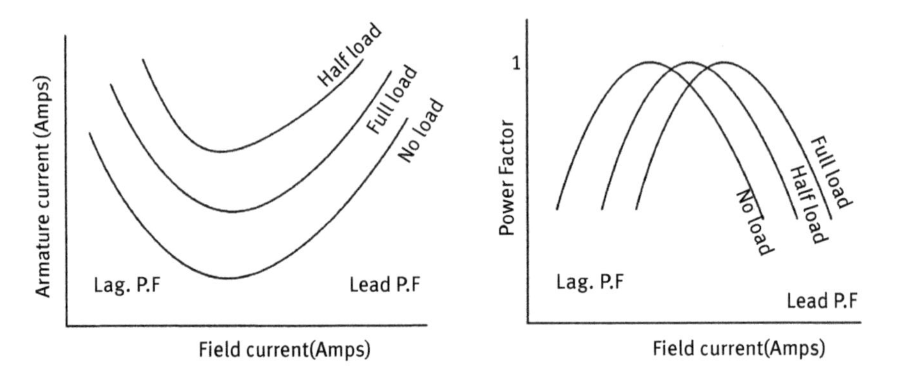

Fig. 6.12: Model graph of "V" and inverted "V" curves of synchronous motor.

6.7.1.6 Practical calculations

This test is conducted on the following rating of the motor under no-load condition and data is tabulated.

Rating of the motor = 5 HP

Current = 4.5 A

Speed = 1,500 rpm

Voltage = 415 V

Observation table

Table 6.3: Practical observations.

S. No.	Supply voltage (V)	Wattmeter reading (W_1)	Wattmeter reading (W_2)	Field current I_f (amp)	Armature current I_a (amp)	$\text{Cos}\phi = \frac{W}{VI_a}$
1	420	200		0	4.4	0.108
2	420	165		0.1	3.7	0.151
3	420	120		0.2	2.6	0.285
4	420	85		0.4	1	0.289
5	420	45	0.5	0.5	0.2	0.535
6	420	20	0.6	0.6	0.1	0.476
7	420	0	0.7	0.7	1	0
8	420	0	0.8	0.8	2.4	0
9	420	10	0.9	0.9	3.8	0.062

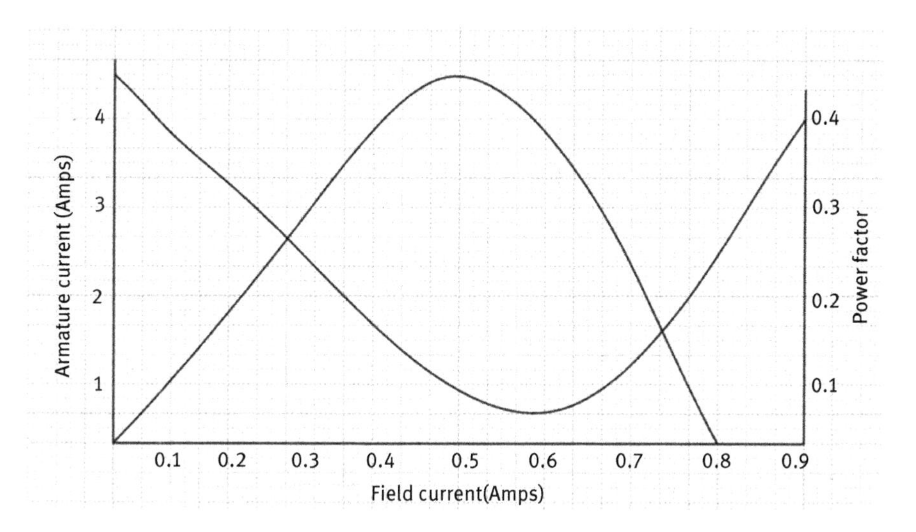

Fig. 6.13: Graph of “V” and inverted “V” curves.

6.7.1.7 Result

Drawn V and inverted V curves of a synchronous motor.

6.8 Viva-Voce questions

1. Define infinite bus.
2. How do you run synchronous generator as motor?
3. Define synchronizing torque.
4. Define torque angle.
5. An over excited synchronous motor is called __________.
6. Explain the effect of field excitation on synchronous motor.
7. Explain the uses of damper bars in synchronous motor.
8. Define Synchronizing torque.
9. Explain the nature of armature reaction in synchronous motor at lagging power factor.
10. Mention the applications of synchronous motor.

6.9 Objective questions

1. __________is used to start synchronous motor.
   (a) Tiny motor
   (b) DC compound motor
   (c) Providing damper winding
   (d) All of the above

2. Hunting occurs in a synchronous motor due to
   (a) continuous changes in load
   (b) running without load
   (c) running under excitation
   (d) All

3. An over excited synchronous motor operating under no-load condition is known as
   (a) booster (b) phase advancer (c) pulse generator (d) regulator

4. If the mechanical load on the synchronous motor is increased three times, then the load (torque) angle increases roughly__________. Assume that the field excitation of the motor is kept constant.
   (a) one third (b) two times (c) three times (d) six times

5. A three-phase synchronous motor will have___________slip rings
   (a) 6 (b) 4 (c) 2 (d) 8

6. The synchronous motor is running at synchronous speed. It is connected to three-phase supply. If Suddenly one phase is disconnected during the operation. Then the synchronous motor will
   (a) reject to start
   (b) stop running
   (c) runs lower than the synchronous speed
   (d) remains run as it is

Answers

1. d, 2. a, 3. b, 4. c, 5. c, 6. a.

6.10 Exercise problems

1. A 2,000 V, 3-phase, 4-pole, Y-connected synchronous motor runs at 1,200 rpm. The excitation is constant and corresponds to an open circuit voltage of 1,500 V. The reactance 3 Ω per phase. Determine the power input, power factor and torque developed for an armature current of 200 A.
2. A 3-phase, 500 V synchronous motor draws a current of 50 A from the supply while driving a c load. The stator is star connected with armature resistance of 0.4 Ω/ph and a synchronous reactance of 4 Ω/ph. Find the power factor at which the motor would operate when the field current is adjusted to give the line emf of 600 V.
3. A 400 V, 50 Hz, 33.7 kW, 3-phase star connected synchronous motor has a full load efficiency of 88%. The synchronous impedance of the motor is (0.2 + j1.6) Ω/ph. If the excitation of the motor is adjusted to give a leading pf of 0.9, Calculate induced emf developed for full load.

Bibliography

[1] Fitzgerald, A. E., Kingsley Jr., C., & Umans, S. Electric Machinery. Tata McGraw-Hill, 2005.
[2] Guru, B. S. & Hiziroglu, H. R. Electric Machinery and Transformers. The Oxford Series in Electrical and Computer Engineering. Oxford University Press, 2000.
[3] Say, M. G. The Performance and Design of Alternating Current Machines. CBS Publishers & Distributors, 2002.
[4] Transformers. Bharat Heavy Electricals Limited. Tata McGraw-Hill, 2003.
[5] Bimbhra, P. S. Electrical Machinery. Khanna Publishers, 1977.
[6] Langsdorf, A. S. Theory of Alternating Current Machinery, Tata McGraw-Hill, 2001.
[7] Theraja, B. L. & Theraja, A. K. A Textbook of Electrical Technology Volume II. S. Chand Publishing, 1959.
[8] www.bhel.com
[9] www.ijems-world.com
[10] www.ieeexplore.ieee.org

https://doi.org/10.1515/9783110682274-007

Index

Armature Reaction 5
Armature Copper Loss 47

Blocked Rotor Test 132
Blondel's Two Reaction Theory 158
Brake Test 49, 59
Brush Contact Loss 47

Circle Diagram 133
Commutators 38
Constant Loses 46
Cumulatively Compound Generator 26
Cumulative Compound Motor 45
Cylindrical Rotor 151

D.C. Generators 1
Differentially Compounded Generator 26
Differentially Compounded Motor 55
Direct Load Test 158

Eddy Current Loss 47
Electro-Magnetic Induction 87

Fleming's Left–Hand Rule 38
Fleming's Right Hand Rule 1
Field Copper Loss 47
Field Test 75

Hopkinson's Test 69
Hysteresis Loss 46

Magnetization Characteristic 21

Power Factor 133

Residual Flux 3
Retardation Test 61
Rotating Magnetic Field 125

Salient Pole Alternators 173
Salient Pole Rotor 150
Scott Connection 114
Self–Excited Generator 3
Separation of Core Losses 103
Shell–Type 89
Slip 125
Slip Test 165
Sumpner's Test 108
Steinmetz Formula 105
Stray Load Loss 48
Swinburne's Test 64
Synchronous Impedance 154
Synchronous Impedance Method 159
Synchronous Reactance 158

Torque 42
Turns Ratio 126

Variable Losses 47
Voltage Regulation 93, 154
Voltmeter Method 114

https://doi.org/10.1515/9783110682274-008